ANALOGUE COMPUTING METHODS

ANALOGUE COMPUTING METHODS

D. WELBOURNE, M.A.

PERGAMON PRESS

OXFORD · LONDON · EDINBURGH · NEW YORK
PARIS · FRANKFURT

Pergamon Press Ltd., Headington Hill Hall, Oxford
4 & 5 Fitzroy Square, London W.1

Pergamon Press (Scotland) Ltd., 2 & 3 Teviot Place, Edinburgh 1

Pergamon Press Inc., 44-01 21st Street, Long Island City, New York 11101

Pergamon Press S.A.R.L., 24 Rue des Ecoles, Paris 5e

Pergamon Press GmbH, Kaiserstrasse 75, Frankfurt-am-Main

First edition 1965
Library of Congress Catalog Card No. 64-8681

Set in 10 on 12 pt. Times and printed in Great Britain by
Blackie & Son Ltd., Glasgow, Scotland

(1713)/65

CONTENTS

Preface ix

Chapter 1 Introduction and Mathematical Techniques x

Introduction 1

Analogue and Digital Computing Machines 2

Historical 3

Mathematical Techniques 5

The Laplace Transform 6

The Transfer Function 7

Electrical Impedance 12

Kirchhoff's Laws 12

Vectors and Scalars 14

Separation of Variables 17

Finite Differences 18

Lagrange Coefficients 20

Matrices 21

Determinants 23

Eigenvalues and Eigenfunctions 24

References 25

Chapter 2 Electronic Analogue Equipment 26

The Amplifier 26

The D.C. Amplifier 26

Basic Relay Controls of the Amplifier 30

Errors of the Simple Amplifier 31

The Chopper Stabilized Amplifier 32
Error Analysis 34
Errors of Summers and Integrators 37
Multipliers 38
The Coefficient Potentiometer 38
The Servo-multiplier 39
Crossed-field Multipliers 41
Hall Effect Multipliers 42
Quarter-squares Multipliers 42
Time-division Multipliers 43
Diode Function Generators and Limiters 44
Resolvers 46
Noise Generators 47
Other Function Generators 48
References 49

Chapter 3 Scaling and Applications 50
Introduction 50
Scaling 51
Summer and Integrator Scaling 52
Time Scaling 54
Multiplier Scaling 55
A Worked Example 57
A Small Problem 61
Estimate of Maximum Values 64
Methods of Solution 65
Some Standard Circuits 67

Chapter 4 Computers and Associated Equipment 72
General Design of a Computer 72
Patchboard 72
Control 73
Measurement 73
Recorders and Readout 74
Overload Indicators and Alarms 74
Maintenance 74
Some Typical Computers 75
The Pace Systems TR-10 75
The Solartron SC-30 75
The Pace Systems 231-R 78

The English Electric Saturn	80
Royal Aircraft Establishment Tridac	84
References	85
Chapter 5 Application of Large Analogue Computers	86
A Nuclear Reactor Simulation—An Example	86
The Physical System	86
Approximations	88
Description of the Equations	88
General Procedure for Preparing a Simulation	97
Simulation Circuits	97
Physical Data	97
Patching and Setting up	98
Procedure on the Machine	98
Programme of Work	99
Some Typical Simulation Results	99
Examples of the Use of Large Analogue Computers	101
The Nuclear Power Industry	101
Aircraft Simulation	102
Missile Systems	103
Process Control	104
Control Engineering	105
References	107
Chapter 6 General Analogue Devices	109
The Differential Analyser	109
Set-up of a Differential Analyser	112
The Incremental Computer	115
Conductive Analogues	117
Resistance Sheets	117
The Electrolytic Tank	119
Resistance Meshes	121
Extensions of the Single Resistance Mesh	124
Resistance–Reactance and Other Meshes	126
Some Special Applications	127
Monte Carlo Methods	127
The Adjoint Method	129
Operational Research	132
References	134
Index	137

PREFACE

THIS book is an attempt to present the field of analogue computation and simulation in a compact and handy form. There is a large number of books, almost all American, which present detailed, thorough, intricate, and all-embracing courses on analogue computing, with the aid of which the student could become a professor of computation, mathematics, control theory and electronics all rolled into one. I have started with the assumption that the reader is intelligent enough to realize the extent and scope of a method without being given large numbers of detailed and unreadable examples; and that, having understood the elements of the subject from the bookwork and examples given, he can use his ability to read specialized works and to chase references.

The reader is assumed to have a knowledge of the calculus, mathematics and physics such as would be adequate to gain admission to a university. A certain amount of mathematics is presented in the first chapter, omission of which will not greatly impair understanding of the remainder of the work. However, in view of the widespread confusion I have found to exist on the topics involved, it was thought well worth including this background.

The electronic equipment of an analogue computer is described in Chapter 2, and its use to solve simple problems shown in Chapter 3, together with the method of scaling. Chapter 4 describes the general layout of a computer with a description of some commercial computers. The use of a large computer is described in Chapter 5 by means of a particular example—a reactor simulation. Types of problem which have been simulated on large computers are described briefly. Finally, Chapter 6 outlines some general analogue devices. The differential analyser is described for historical reasons and because of the insight it has been found to give some people into the solution of equations by simulation methods. The methods used in the incremental computer (or digital differential analyser) are explained briefly. Conductive analogues and some other methods are described, and, finally, some special techniques, which have been used in connexion with electronic analogue computing, are given.

I am very greatly indebted to the authors of many works, notably W. J. Karplus's excellent book *Analogue Simulation in the Solution of Field Problems*, and also to A. E. Rogers and T. W. Connoly, *Analog Computation in Engineering Design*, both published by McGraw-Hill. *Analog Computation*, by A. C. Jackson (McGraw-Hill), is an extremely useful book. The bibliography included at the end of the book will be helpful to those who would dig deeper into the subject than this book attempts.

I would like to express my gratitude to Percy Hammond for asking me to write this book, and to Dr. T. O. Jeffries, under whom I was working at the time, who gave me much useful advice and encouragement with the text.

My thanks are due to Solartron and to Pace systems for permission to publish description of their computers, and to the English Electric Co. for permission to write this book.

Finally, I would like to record my thanks to Miss I. London who typed the manuscript.

chapter 1

INTRODUCTION AND MATHEMATICAL TECHNIQUES

INTRODUCTION

Geometry is one of the oldest branches of mathematics. It is supposed to have arisen out of the desire of the Ancient Egyptians to make accurate maps in order to settle disputes over land-holders' boundaries. Their maps were analogues of the spatial relationships of neighbouring areas of land, and the analogue was available in a convenient form after the floods of the Nile had removed the boundary markers.

Today every draughtsman is familiar with the geometrical methods first thought out by the Greeks and Egyptians. With their aid he can lay out in a convenient and readily changed form the design of a complex machine or building to high accuracy.

Man does much of his thinking by drawing analogies between something which is partly understood and something which is more fully understood or which is more readily handled. The physicist explains some aspects of the behaviour of atoms and electrons by comparing them to billiard balls, the barrister points to a precedent, a case similar to the one in question, to help decide the answer to his case, and the theatre seeks to clarify our personal relationships by acting a similitude of real life.

Many devices have been thought up to help in the understanding of the problems of engineering, physics and mathematics.

Mathematics and physics are concerned, among other things, with finding models, expressed as relationships between numbers, for the world as our senses perceive it. And the engineer applies these models, which are analogues of the physical world.

This book is intended to give an outline of some of the many models and analogues that have been produced to solve physical problems for the engineer, and, more particularly, to show how to use and programme the modern electronic analogue computer.

ANALOGUE AND DIGITAL COMPUTING MACHINES

Basically the digital computer does sums, sums that most people could do by the age of 10 or so, takes two or three simple types of decision, has an enormous memory, and works at very high speed. It does arithmetic.

The analogue computer is any arrangement which can be made to behave in as near the same way as possible to some real system of some kind. Most, if not all, digital computers are essentially serial in operation (even those known as parallel machines), that is, the numbers stored in the machine appear in some central arithmetic unit one after another and are then added to or compared with some other number before being sent back to the memory store. At one time, only one "sum" is being done on the whole set of numbers in the computation. By the use of very small computing times, of the order of a few microseconds, and stores of very large capacity, very large and complex problems can be handled.

Analogue machines are essentially parallel devices. All the numbers describing the problem, represented as voltages or shaft positions or as pressures, flows or distances, are all processed simultaneously. Additions, multiplications and integrations are all proceeding at the same time. A widely used device is the wind tunnel where the physical variables such as air speed and force are represented by air speed and force, but are scaled down suitably to give a representation of a large structure by a small model.

The digital machine is capable of great accuracy and speed, but it needs a skilled mathematician to programme it. Great care must

be taken to avoid giving the machine impossible situations, or getting it going round in circles trying to find an answer to a faulty logical situation. On an analogue, an engineer can quickly get the feel of a design and alter parameters and see what happens. He will often operate the machine himself.

The experts say that any problem can be solved on a digital machine. But it may take far longer to programme, test the programme, and run off the solutions than to run an analogue machine programme. Nuclear reactor studies, rocket systems, and problems involving many non-linearities can frequently be solved most economically on an analogue computer. It is possible to include parts of the physical system under test in the analogue set-up in some cases.

The accuracy of an analogue machine is seldom better than one part in 1000. This is better than the accuracy of the physical data of many problems. But this is sometimes too low, and the additional cost of a digital programme becomes necessary. A fairly detailed costing of the computation of some radiation integrals involving Bessel functions has been done by Hansen. The analogue solution, accurate to two figures, took 2 hours to programme, 50 minutes to compute the integrals required, and was costed at \$53. The digital programme took 2 weeks to write and 50 minutes on the machine, and was costed at \$1377. This was much the same as the cost of a hand calculation.

However, at the present time, a large analogue problem can tie up a computer for weeks and months, and no other problem can be put on the machine during this time since the constants of the problem are represented as potentiometer settings. This can be overcome if some automatic means is used for setting up the problem, but these are not yet widely available. In contrast, a digital computer reads in its programme and data, computes, and if everything goes well, prints out the answers and is clear for the next operator in 10 minutes, an hour, or in extreme cases a night's work.

HISTORICAL

The first device described for solving differential equations was an integrating machine developed from the principle of the plani-

meter by Prof. James Thomson, Lord Kelvin's brother. Its construction and use was described in a paper to the Royal Society in 1875. A ball is allowed to roll on a disc, and the number of rotations is transmitted to a roller. This is described in Chapter 5. The roller then gives a total rotation proportional to the integral of the distance of the ball from the centre of the disc with respect to the rotation of the disc.

Although Lord Kelvin indicates how the device can be used for the solution of differential equations, and quotes problems of the vibration of stretched chords, water waves in channels, heat transfer and the motion of the tides as being suitable, it was not used until Vannevar Bush re-invented an integrator at M.I.T. in America, and built a six-integrator machine in the 1920's. This and a later machine were used in the solution of problems of synchronous motor control and stability, and emitting filaments, among many others. During and after the Second World War, various machines were built, ranging from a two-integrator Meccano machine built by Prof. Hartree, accurate to $1\frac{1}{2}$ per cent, to a thirty-integrator machine, accurate to 1 part in 10,000, built at M.I.T. with electric couplings and drive, and a paper tape input and output. These machines were widely used for such things as control and vibration problems.

After the Second World War, valves were good enough and cheap enough for the construction of high-gain d.c. amplifiers of fairly good stability to be possible. Several computers were built using these amplifiers to perform summing and integrating functions in a manner that will be described later. Careful design enabled reliable computers of up to about 100 amplifiers to be built, but the difficulty of amplifier output drift was only overcome by the widespread introduction of the chopper relay stabilization technique developed by E. A. Goldberg in about 1950. Since then, a great many computers of up to about 100 amplifiers and some of as many as 1500 amplifiers have been built. These are reliable, accurate to about 0·1 per cent, and have been used for an enormous variety of problems in, for example, control, heat transfer, chemical engineering, missile system development, and nuclear reactor studies.

The development of analogue computers has been parallel to and in some ways complementary to the development of digital

computers. A great deal of nonsense and rivalry has appeared, where one method has been acclaimed as so superior to the other that the other is considered to be outdated. But each method of approach to the solution of a problem will have its uses. Analogue solutions are often cheaper, quicker, simpler to programme, but far less accurate than digital computer solutions; but there are many problems which it is not reasonable to attempt to put on to an analogue machine, such as the solution of large sets of linear equations, the accurate evaluation of integrals, or the compilation of tables, e.g. logarithms. The fact that both methods have their own uses is shown by the many firms which have found it economic to install both analogue and digital computing systems. Often one will feed data to the other, and particular cases can be solved on both, enabling cross-checking of results to be carried out.

MATHEMATICAL TECHNIQUES

The programmer of an analogue computer may be given sets of equations to solve which involve unfamiliar notations and mathematical language, although the equations themselves do not involve any really difficult mathematics. The writer has met many users of analogue machines who have been puzzled by the mathematical language of the problem which they are in fact solving. Therefore, the following is provided as a sort of mathematical "phrase book", without any attempt at a serious study of the "grammar" of the mathematics involved. It is hoped that the trained mathematicians will forgive me for the sketchy way the subjects are covered. I know that the material of this chapter has proved helpful to my engineering colleagues, but I fear the wrath of my mathematical lecturers and teachers.

On this basis, a short outline of the Laplace transform is given and of the use of the operator p (or s). The operator del $\equiv \nabla$ is explained briefly, and types of equation where it appears are mentioned. A short example illustrates the method of separation of variables. Finite difference forms of differentials and Lagrange coefficients are given. The notation of matrices is explained, and the existence of eigenvalues mentioned, together with some remarks on eigenfunctions.

THE LAPLACE TRANSFORM

The Laplace transformation provides a technique for solving linear differential equations by algebraic methods. A short outline is given here. The student will find references to fuller treatments at the end of the chapter.

We define the transform $\bar{f}(p)$ of a function $f(t)$ by the relation

$$\bar{f}(p) = \int_0^\infty f(t)\,e^{-pt}\,dt \qquad (1)$$

This is subject to mathematical conditions concerning the limits and validity of the integration, which are satisfied in the majority of cases ordinarily encountered. Two theorems are readily derived:

If $\bar{f}(p)$ is the Laplace transform of $f(t)$,

(1) $\bar{f}(p+a)$ is the transform of $f(t)e^{-at}$;

(2) $(d/dt)f(t)$ transforms to $p\bar{f}(p)-y_0$, where $y_0 = f(0+)$,† and applying this result twice,

$$(d^2/dt^2)f(t) \text{ transforms to } p^2\bar{f}(p)-py_0-y_1,$$

TABLE 1

Function		Laplace transform
Impulse	$\delta(t)$	1
Step	$H(t)$	$1/p$
Ramp	t	$1/p^2$
	$\dfrac{t^n}{n}$	$\dfrac{1}{p^{n+1}}$
	e^{-at}	$\dfrac{1}{p+a}$
	$t^n e^{-at}$	$\dfrac{n!}{(p+a)^n}$
	$\sin \omega t$	$\dfrac{\omega}{\omega^2+p^2}$
	$\dfrac{df}{dt}$	$p\bar{f}(p) - f(0+)$

† The notation $f(0+)$ denotes the limiting value of $f(t)$ as t tends to zero, where t is always greater than zero.

where y_1 and y_0 are the values of $\mathrm{d}f/\mathrm{d}t$ and f at time $t=0+$. It can easily be shown that the transform of a sum of two functions is the sum of the transforms, and that the transform of $Kf(t)$ is $K\bar{f}(p)$, where K is a constant.

Integration of equation (1) leads to the transforms shown in Table 1.

Exercise

Prove the above theorems by integration from the definition of the transformation.

Solution of a Simple Equation

Consider the equation $a\ddot{x}+b\dot{x}+cx=kf(t)$. Take the Laplace transform of both sides;

Then $$(ap^2+bp+c)\bar{x} = k\bar{f}(p)+(ap+b)x_0+ax_1$$

thus $$\bar{x}(p) = \frac{k\bar{f}(p)+(ap+b)x_0+ax_1}{ap^2+bp+c}.$$

By expanding in partial fractions and the use of the table of transforms, the function $x(t)$ can be found given the initial conditions and the function $kf(t)$. This is a simple matter for this equation, and the solution can be found as quickly by other means. But the technique is as easy to apply for simultaneous linear differential equations, or equations of high order, the difficulty lying in the resolution into partial fractions of the expression equivalent to the r.h.s. in the equation above.

Some people use the symbol s for the Laplace operator rather than p. The writer can detect no majority view, and no reason to be partisan. Since the symbols are interchangeable, p has been selected on arbitrary grounds.

THE TRANSFER FUNCTION

If a system has input $f(t)$, output $x(t)$, then the function

$$G(p) = \frac{\bar{x}(p)}{\bar{f}(p)}$$

is called the transfer function of the system. Note that it can also be obtained by replacing the differential operator $\mathrm{d}/\mathrm{d}t$ by the symbol p in the differential equation of a linear system.

Sinusoidal Inputs

For a linear system with sinusoidal input, all the system variables will be sinusoidal, differing only in amplitude and phase from the input.

If $$F(t) = A\,e^{+j\omega t} = (a+jb)\,e^{+j\omega t},$$

where a and b are real numbers, then if

$$A = a^2+b^2 \quad \text{and} \quad \tan\phi = b/a,$$

it is easily shown that

$$F(t) = A\cos(\omega t+\phi)+jA\sin(\omega t+\phi).$$

So we can represent a sinusoid in phase and magnitude, in a linear system, by a complex number of the form $A\,e^{j\omega t}$. Any algebra involved in the system can be performed on this number. Taking real parts at any stage gives the same result as would have been obtained if the solution had been worked through by operating on the real part alone. Now

$$\frac{d}{dt}(A\,e^{+j\omega t}) = j\omega A\,e^{+j\omega t}.$$

Therefore the operation of differentiation is formally identical to multiplication by $j\omega$. It is convenient to write $j\omega = p$, so $d/dt \equiv p = j\omega$.

The differential equation describing a linear system can therefore be written in the form

$$a_n p^n X + a_{n-1} p^{n-1} X + \ldots + a_1 pX + a_0 X$$
$$= b_m p^m F + b_{m-1} p^{m-1} F + \ldots + b_1 pF + b_0 F,$$

where F is the input and X the output sinusoid of the system.

The transfer function defined as the ratio of the Laplace transforms of output and input would give exactly the same expression except that the complex numbers X and F representing the output and input functions x and f would be replaced by the transforms $\bar{x}$ and $\bar{f}$. The term "transfer function" is used to cover both expressions. Care should be taken in interpreting this function. When it is used in the sinusoidal sense, the modulus of $G(j\omega)$ will give the gain, and the ratio of its imaginary to real parts the tangent of the phase. When used in its Laplace transform sense, it is an operator, operating on the transform of the input to give the transform of the output.

However, since the two expressions are formally identical, we can obtain the gain and phase of the transfer function simply by replacing the symbol p by $j\omega$ and taking the modulus and argument as before.

Decibels

The decibel is the measure of the logarithm of the ratio between two quantities which measure powers. The relationship in decibels (abbreviated to dB) between powers P_0 and P_1 is defined by

$$\text{number of decibels} = 10\log_{10}\left(\frac{P_0}{P_1}\right).$$

Thus, if the input and output voltages of a system are V_1 and V_0, since the power is proportional to the voltage squared, the gain in decibels of the system is

$$\text{gain in decibels} = 20\log_{10}\left(\frac{V_0}{V_1}\right).$$

Now, consider the simple lag, with transfer function

$$G(p) = 1/(1+pT).$$

The gain is the modulus of this, with p replaced by $j\omega$;

$$\gamma(\omega) = \left|\frac{1}{1+j\omega T}\right| = \sqrt{\left(\frac{1}{1+\omega^2T^2}\right)}$$

So

$$\gamma \text{ in decibels} = 20\log\sqrt{\left(\frac{1}{1+\omega^2T^2}\right)}$$
$$= -10\log(1+\omega^2T^2).$$

When ω is small, the gain is $= 0$ dB.

When ω is large, the gain is $\doteqdot -10\log(\omega^2T^2)$,

$$\doteqdot -20\log\omega T.$$

Thus for large values of ω the gain decreases by 20 dB for each decade (factor of 10) in frequency; this is equivalent to 6 dB for each octave (doubling in frequency), since $\log 2 = 0{\cdot}3$ very nearly.

Similarly, the phase is given by $\phi(\omega) = -\tan^{-1}\omega T$. Now, on

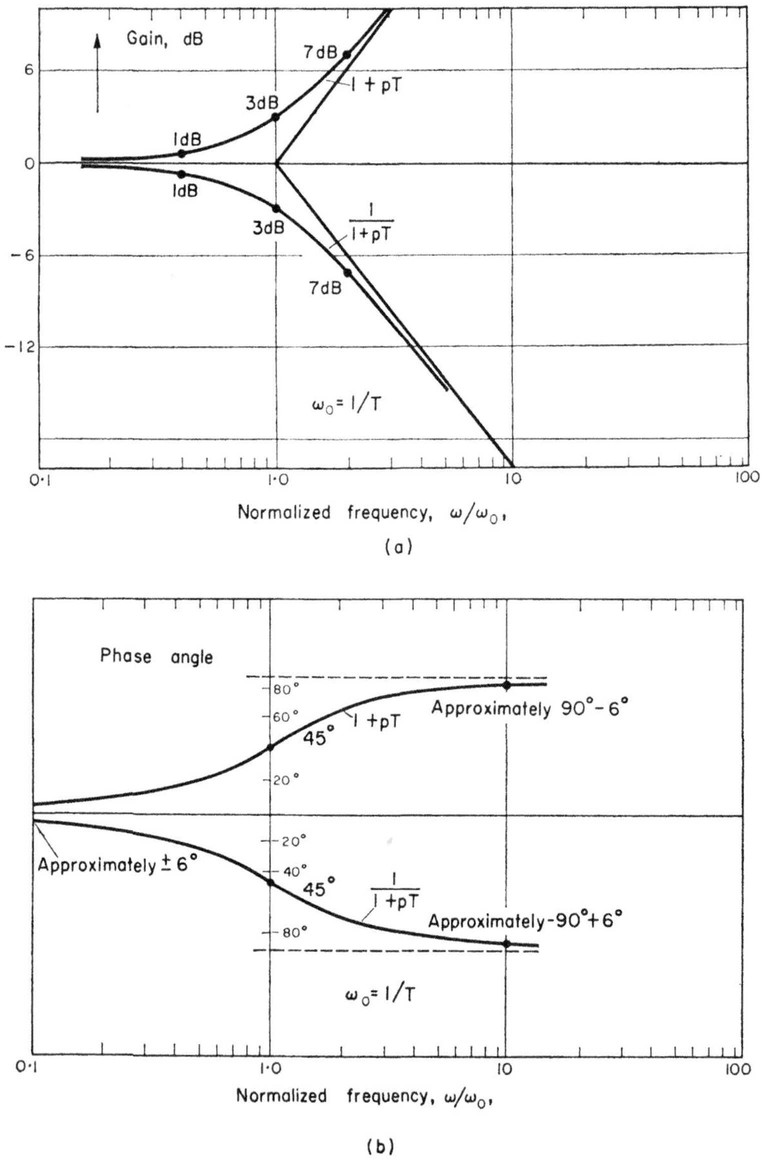
Gain, dB
6
0
-6
-12
7dB
3dB
1dB
1 + pT
1dB
3dB
1/(1+pT)
7dB
ω₀ = 1/T
0·1
1·0
10
100
Normalized frequency, ω/ω₀,
(a)
Phase angle
80°
60°
1 +pT
45°
Approximately 90°-6°
20°
20°
Approximately ± 6°
40°
45°
1/(1 +pT)
Approximately-90°+6°
80°
ω₀ = 1/T
0·1
1·0
10
100
Normalized frequency, ω/ω₀,
(b)

FIG. 1.

a logarithmic scale of ω, the points corresponding to $\omega = \alpha/T$ and $\omega = 1/\alpha T$ are symmetrical about the point $\omega = 1/T$; also, since at the point $\omega = 1/T$, $\phi = -45^\circ$, and since

$$\tan^{-1}\frac{\alpha}{T}.T + \tan^{-1}\frac{1}{\alpha T}.T = 90^\circ,$$

the curve of phase against frequency, with frequency on a logarithmic scale, is symmetrical about 45° and $\omega = 1/T$, as shown in Fig. 1.

EXERCISES

1. Calculate the gain and phase of the transfer functions $G(p) = p$, $1/p$, $1 + pT$, and $1/(1 + pT)$ as a function of frequency. Show that the gain of the third and fourth can be approximated by appropriately placed asymptotes with slope 6 dB/octave as shown in Fig. 1a.
2. A network has transfer function

$$G(p) = \frac{p}{(1 + p)(1 + 0{\cdot}1p)}.$$

What is its response to a ramp input? By using the asymptote method outlined in Ex. 1, sketch its frequency response on a log-linear scale.

The p-plane

We may factorize a transfer function $G(p)$ to give an expression of the form

$$G(p) = \frac{(\beta_1 - p)(\beta_2 - p)\ldots}{(\alpha_1 - p)(\alpha_2 - p)\ldots},$$

where the α's and β's are real, or complex pairs, or may be repeated.

Then, if $p = \alpha + j\omega$, where α and ω are real, we may mark on the Argand diagram of the complex numbers the poles of $G(p)$, where $p = \alpha_1, \alpha_2, \ldots$, and the zeros, where $p = \beta_1, \beta_2, \ldots$ (A pole of $G(p)$ is a point where its value is indeterminately large.) Suppose we have a pair of poles at $a + jb$, $a - jb$; in the partial fraction expansion of the response there will therefore be a term of the form

$$\frac{A}{p-(a+jb)} + \frac{B}{p-(a-jb)}.$$

Reference to the table of transforms shows that the output contains terms of the form $Ae^{at}.e^{jbt}+Be^{at}.e^{-jbt}$. Thus, if a is greater than 0, this represents an unstable response. So poles in the right-hand half-plane mean a diverging response.

Details of the techniques developed for the investigation of transfer functions are given in any textbook on control theory.

Exercise

Investigate the behaviour of the system with transfer function

$$G(p) = \frac{1}{p^2 + 2\zeta\omega_n p + \omega_n^2}$$

as the damping coefficient ζ varies from $+\infty$ to $-\infty$. A thorough study of this function will give a great deal of insight into the behaviour of many physical systems.

Electrical Impedance

Ohm's law is well known in the form $E = IR$. If in the differential equations for the flow of current through a capacitor and an inductance, p is written for d/dt, we obtain $E = (1/Cp).I$ and $E = Lp.I$. The quantities $1/Cp$ and Lp are called the impedances of a capacitor and an inductance respectively in operational form. Reference to the differential equations quickly show that the rules for combining impedances in series and parallel are the same as those for combining resistors.

Exercise

What is the impedance of an inductance, a capacitor and a resistance in series, and in parallel? Show that the impedances and transfer functions of the networks given in Fig. 2 are those given in the table. Derive other transfer functions from combinations of various impedances.

Kirchhoff's Laws

(1) At any junction of impedances, called a node, the total inflow of current equals the total outflow.

(2) Round any closed path in a circuit, called a loop, the sum of the currents in each element times its impedance is equal to the voltage drop round the loop.

These follow from the conservation of electric charge at the

(a)

Network	Impedance
	R
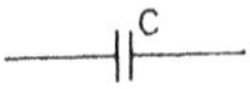	$\frac{1}{pC}$
	pL
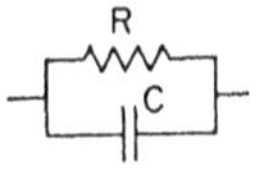	$\frac{R}{1 + pCR}$
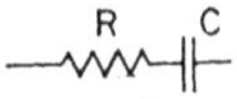	$\frac{1 + pCR}{pC}$
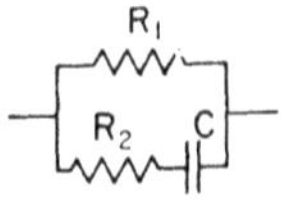	$\frac{R_1(1 + pCR_2)}{1 + pC(R_1 + R_2)}$
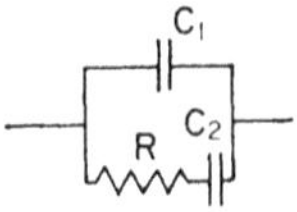	$\frac{(1 + pC_2R)(C_1 + C_2)}{p\left(1 + p\frac{C_1C_2}{C_1 + C_2}R\right)}$

(b)

Network	Transfer Function
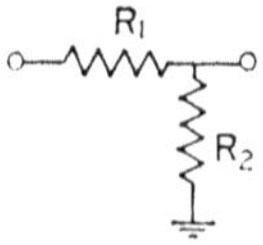	$\frac{R_2}{R_1 + R_2}$
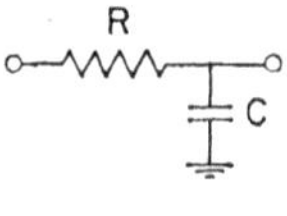	$\frac{1}{1 + pCR}$
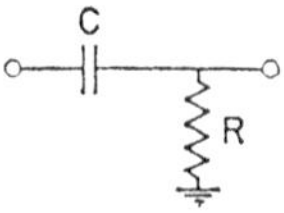	$\frac{pCR}{1 + pCR}$
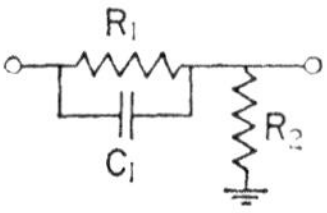	$\frac{R_2}{R_1 + R_2} \cdot \frac{1 + pC_1R_1}{1 + pC_1\frac{R_1R_2}{R_1 + R_2}}$
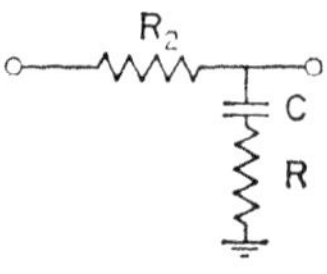	$\frac{1 + pCR_1}{1 + pC(R_1 + R_2)}$

FIG. 2.

node, and the application of Ohm's law to each element of the loop.

VECTORS AND SCALARS

It will be recalled that many physical variables have a direction as well as a magnitude associated with them. Force and velocity are familiar examples of vectors. Vectors are combined by the parallelogram rule, normally introduced as the parallelogram of forces. A vector is distinguished in print by bold lettering.

A scalar quantity is one such as mass or volume which combines by simple addition. Two pounds plus three pounds makes five pounds. This is contrasted with vectors, which combine thus: a ship heading 30° west of north at 10 knots in a river flowing due east at 5 knots will move due north at $5\sqrt{3}$ knots.

A vector can be expressed as a function of position; the force due to gravity near the earth is directed towards the earth and is inversely proportional to the square of the distance from the earth's centre. Similarly, a scalar quantity can be expressed as a function of position—for example, the temperature of the air in a room as a function of the position in the room.

The mathematical use of vectors has enabled many branches of physics to be described in a compact notation. Differentiation of a vector takes a special form, and various related operations of differentiation are found convenient. These relate to the important differential equations such as Laplace's equation, the diffusion equation and the wave equation.

The Operator ∇ (Del)

"Del" is a differential operator defined by

$$\nabla \equiv \mathbf{i}\frac{\partial}{\partial x}+\mathbf{j}\frac{\partial}{\partial y}+\mathbf{k}\frac{\partial}{\partial z}$$

where **i**, **j**, **k**, are unit vectors in the x-, y-, and z-directions.

If $\phi(x,y,z)$ is a scalar function defined in a coordinate system (x,y,z), then the vector

$$\operatorname{grad}\phi = \nabla\phi = \mathbf{i}\frac{\partial\phi}{\partial x}+\mathbf{j}\frac{\partial\phi}{\partial y}+\mathbf{k}\frac{\partial\phi}{\partial z}$$

is called the gradient of ϕ. It is analogous to the slope of a function $f(x)$ that is, $f'(x) = (\mathrm{d}/\mathrm{d}x)f(x)$. The three components of the slope are given by the derivatives in the directions of the three axes.

If $\mathbf{a}(x,y,z) = (a_x, a_y, a_z)$ is a vector defined in the system of coordinates, then the scalar

$$\operatorname{div} \mathbf{a} = \nabla . \mathbf{a} = \frac{\partial a_x}{\partial x} + \frac{\partial a_y}{\partial y} + \frac{\partial a_z}{\partial z}$$
$$= \left(\mathbf{i} \frac{\partial}{\partial x} + \mathbf{j} \frac{\partial}{\partial y} + \mathbf{k} \frac{\partial}{\partial z} \right)(\mathbf{i}a_x + \mathbf{j}a_y + \mathbf{k}a_z)$$

is called the divergence of the vector **a**. Consider a small sphere around the point (x,y,z); then div **a** measures the flux of **a** leaving the point and crcssing this sphere, in the limit, as the volume of the sphere shrinks to zero. We will show how the first term arises.

Consider a small box, sides δx, δy, δz, with the point (x,y,z) as one corner. Then the flux of **a** entering across the face perpendicular to the x-axis at the point (x,y,z) is $a_x \delta y . \delta z$. The flux leaving the parallel face at $x + \delta x$ is then

$$\left(a_x + \frac{\partial a_x}{\partial x} \delta x \right) \delta y . \delta z.$$

Thus the net flux leaving the box is $(\partial a_x / \partial x) . \delta x . \delta y . \delta z$ in the x-direction. This immediately gives the first term of div **a** as the volume $\delta x \delta y \delta z$ shrinks to zero, per unit volume.

Finally, a vector is defined by

$$\operatorname{curl} \mathbf{a} = \nabla \times \mathbf{a} = \left(\mathbf{i} \frac{\partial}{\partial x} + \mathbf{j} \frac{\partial}{\partial y} + \mathbf{k} \frac{\partial}{\partial z} \right) \times (\mathbf{i}a_x + \mathbf{j}a_y + \mathbf{k}a_z)$$
$$= \begin{vmatrix} \mathbf{i} & \mathbf{j} & \mathbf{k} \\ \frac{\partial}{\partial x}, & \frac{\partial}{\partial y}, & \frac{\partial}{\partial z} \\ a_x, & a_y, & a_z \end{vmatrix}$$
$$= \mathbf{i} \left(\frac{\partial a_z}{\partial y} - \frac{\partial a_y}{\partial z} \right) + \mathbf{j} \left(\frac{\partial a_x}{\partial z} - \frac{\partial a_z}{\partial x} \right) + \mathbf{k} \left(\frac{\partial a_y}{\partial x} - \frac{\partial a_x}{\partial y} \right).$$

This vector is a measure of the work done in passing round the point (x,y,z) in a small loop perpendicular to the x-, y-, and z-axes in turn.

Reference should be made to books on vector methods for a full treatment. (D. E. Rutherford, *Vector Methods*, Oliver & Boyd.)

It is frequently the case that we can define a potential function $\phi(x,y,z)$ such that the force experienced by a test object can be written as $\mathbf{a} = -\mathrm{grad}\,\phi$. (For example, the electrostatic force on a test charge a distance x from a charge e is given by $e/x^2 = (-\mathrm{d}/\mathrm{d}x)(e/x)$, and e/x is the potential function in one dimension.) Gauss's theorem shows that the flux of this force leaving a surface around a point is zero unless the surface encloses a "source", for example, an electric charge, and results in the general law expressed by Poisson's equation:

$$-\nabla . \mathbf{a} = \nabla^2\phi = 4\pi\rho,$$

where ρ is the source density. This gives Laplace's equation:

$$\nabla^2\phi = 0$$

in a region without any sources.

These two equations are of the utmost importance in engineering, and describe the behaviour of such things as fluid flow, electrostatics, elasticity, and an immense variety of other problems.

For reference, the form of ∇ and of ∇^2 is given here in Cartesian coordinates, cylindrical polar coordinates, and spherical coordinates.

$$\nabla \equiv \left(\frac{\partial}{\partial x}, \frac{\partial}{\partial y}, \frac{\partial}{\partial z}\right) \qquad (x, y, z)$$

$$\equiv \left(\frac{\partial}{\partial r}, \frac{1}{r}\frac{\partial}{\partial \theta}, \frac{1}{r\sin\partial}\frac{\partial}{\partial \psi}\right) \qquad (r, \theta, \psi)$$

$$\equiv \left(\frac{\partial}{\partial r}, \frac{1}{r}\frac{\partial}{\partial \theta}, \frac{\partial}{\partial z}\right) \qquad (r, \theta, z)$$

$$\nabla^2 \equiv \frac{\partial^2}{\partial x^2} + \frac{\partial^2}{\partial y^2} + \frac{\partial^2}{\partial z^2} \qquad (x, y, z)$$

$$\equiv \left(\frac{\partial^2}{\partial r^2} + \frac{2}{r}\frac{\partial}{\partial r}\right) + \left(\frac{1}{r^2}\frac{\partial^2}{\partial \theta^2} + \frac{\cot\theta}{r^2}\frac{\partial}{\partial \theta}\right) + \frac{1}{r^2\sin^2\partial}\frac{\partial^2}{\partial \psi^2} \qquad (r, \theta, \psi)$$

$$\equiv \left(\frac{\partial^2}{\partial r^2} + \frac{1}{r}\frac{\partial}{\partial r}\right) + \left(\frac{1}{r^2}\frac{\partial^2}{\partial \theta^2}\right) + \frac{\partial^2}{\partial z^2} \qquad (r, \theta, z).$$

Note the different radius derivatives in ∇^2 for (r,θ,ψ) and (r,θ,z) coordinates.

SEPARATION OF VARIABLES

Consider a two-dimensional form of the wave equation

$$\frac{\partial^2\phi}{\partial r^2}+\frac{1}{r}\frac{\partial\phi}{\partial r}+\frac{1}{r^2}\frac{\partial^2\phi}{\partial\theta^2}=\frac{1}{c^2}\frac{\partial^2\phi}{\partial t^2}.$$

This could result from a problem with cylindrical symmetry, and with no variations in the z-direction. Suppose we may write

$$\phi = R(r)\,\Theta(\theta)\,T(t).$$

Then $$\Theta T\left(\frac{d^2R}{dr^2}+\frac{1}{r}\frac{dR}{dr}\right)+RT\left(\frac{1}{r^2}\frac{d^2\Theta}{d\theta^2}\right)=\frac{R\Theta}{c^2}\frac{d^2T}{dt^2}.$$

Therefore $$\frac{1}{R}\left(\frac{d^2R}{dr}+\frac{1}{r}\frac{dR}{dr}\right)+\frac{1}{\Theta}\left(\frac{1}{r^2}\frac{d^2\Theta}{d\theta^2}\right)=\frac{1}{Tc^2}\frac{d^2T}{dt^2}.$$

The l.h.s. is independent of t, so we may equate it to a constant, $-n^2$.

Therefore $$\frac{d^2T}{dt^2}+n^2c^2T=0 \qquad (1)$$

and $$\frac{r^2}{R}\frac{d^2R}{dr^2}+\frac{r}{R}\frac{dR}{dr}+n^2r^2=-\frac{1}{\Theta}\frac{d^2\Theta}{d\theta^2}.$$

Again, the l.h.s. is independent of θ, so we write

$$r^2\frac{d^2R}{dr^2}+r\frac{dR}{dr}+(n^2r^2-k^2)R=0 \qquad (2)$$

and $$\frac{d^2\Theta}{d\theta^2}+k^2\Theta=0. \qquad (3)$$

The solution of (1) is $T = A\cos cnt$,

(3) is $\Theta = B\cos k\theta$,

(2) is a Bessel function of order k,

$$R = CJ_k(nr)+DY_k(nr).$$

Therefore $$\phi=(A\cos cnt)(B\cos k\theta)\left[CJ_n(nr)+DY_k(nr)\right].$$

It is frequently useful to separate the variables in a differential equation in this way. In the above example, suppose we were looking for solutions, periodic in time, in a cylinder of radius a, where ϕ is to be finite, single valued, and zero at the boundary at radius a.

Then $$\phi_{n,k} = A\cos nct \cos k\theta J_k(nr),$$

where n is to be chosen to give $J_k(na) = 0$; this has first, second, ..., mth zeros corresponding to values of n which can be found from tables, and we may describe the solution completely by the two integers k and $n = m$.

The reader should not shy away from the perhaps unfamiliar words "Bessel functions". Bessel functions are a set of wave-like functions which tend to appear in engineering and physics when a problem has cylindrical symmetry. They are tabulated (like sine and tangent) in books of tables.

It is often the case that, as with a Fourier series, we can build up a solution to fit certain boundary conditions from a linear combination, that is, a sum of multiples, of such functions. These functions are called the eigenfunctions or modes, and the numbers corresponding to m or n, and k in the above are called eigenvalues, or characteristic values of the problem. A simple example is a vibrating string; Pythagoras showed that the musical notes have a relationship described by the first, second, third, . . . , modes of vibration of a stretched string. Here the modes or eigenfunctions take the form $A\sin(nx)$ and the eigenvalues m the value of n.

FINITE DIFFERENCES

The definition of the derivative $f'(x)$ of the function $f(x)$ is

$$f'(x) = \lim_{\delta x \to 0} \frac{f(x+\delta x) - f(x)}{\delta x} = \frac{\mathrm{d}f(x)}{\mathrm{d}x}.$$

In analogue work it is frequently the case that the differential equations describing the problem involve two or more independent variables. Consider, for example, the flow of heat through a plate heated at one edge. The flow of heat is proportional to the temperature gradient, so this gradient must be written in an approximate form by going back to the definition written above. The plate is divided up into a mesh of points, and the temperature gradient at a point written in terms of the difference of temperature between two points and the distance between them.

Thus, if x_0, x_1, x_2 are three successive values of x, a distance h apart,

$$\frac{\mathrm{d}f_1}{\mathrm{d}x} = \frac{(x_1)-f(x_0)}{h} \qquad \text{backward difference.}$$

$$= \frac{f(x_2)-f(x_1)}{h} \qquad \text{forward difference.}$$

$$= \frac{f(x_2)-f(x_0)}{2h} \qquad \text{central difference.}$$

Hence

$$\frac{\mathrm{d}^2 f_1}{\mathrm{d}x^2} = \frac{1}{h}\left(\frac{f(x_2)- \ (x_1)}{h} - \frac{f(x_1)-f(x_0)}{h}\right)$$

$$= \frac{f(x_2)+f(x_0)-2f(x_1)}{h^2}.$$

It is readily seen that, in two dimensions, if f_1, f_2, f_3, f_4 are the values of $f(x,y)$ at the points 1, 2, 3, 4 shown in Fig. 3, then

$$\frac{\partial^2 f_0}{\partial x^2}+\frac{\partial^2 f_0}{\partial y^2} = \frac{1}{h^2}(f_1+f_2+f_3+f_4-4f_0) = f''_0$$

And, in three dimensions,

$$f''_0 = \frac{1}{h^2}(f_1+f_2+f_3+f_4+f_5+f_6-6f_0).$$

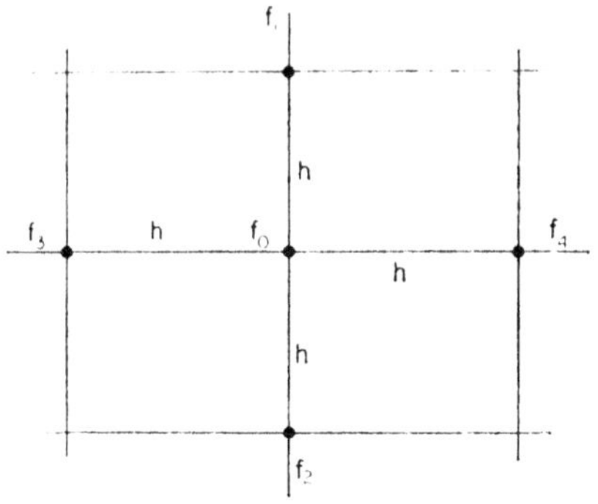

FIG. 3.

A more sophisticated approach is to expand the function as a Taylor series about the point. Thus, in the lattice shown in Fig. 4 we have

$$f_2 = f_0 + a_2 \frac{\partial f_0}{\partial x} + \frac{a_2^2}{2!}\frac{\partial^2 f_0}{\partial x^2} + \frac{a_2^3}{3!}\frac{\partial^3 f_0}{\partial x^3} + \ldots$$

$$f_1 = f_0 - a_1 \frac{\partial f_0}{\partial x} + \frac{a_1^2}{2!}\frac{\partial^2 f_0}{\partial x^2} - \frac{a_1^3}{3!}\frac{\partial^3 f_0}{\partial x^3} + \ldots$$

Therefore $$\frac{\partial^2 f_0}{\partial x^2} = \frac{a_1 f_2 + a_2 f_1 - (a_1 + a_2) f_0}{\frac{1}{2} . a_1 a_2 (a_1 + a_2)} - \frac{(a_2^2 - a_1^2)}{3(a_1 + a_2)}\frac{\partial^3 f}{\partial x^3}.$$

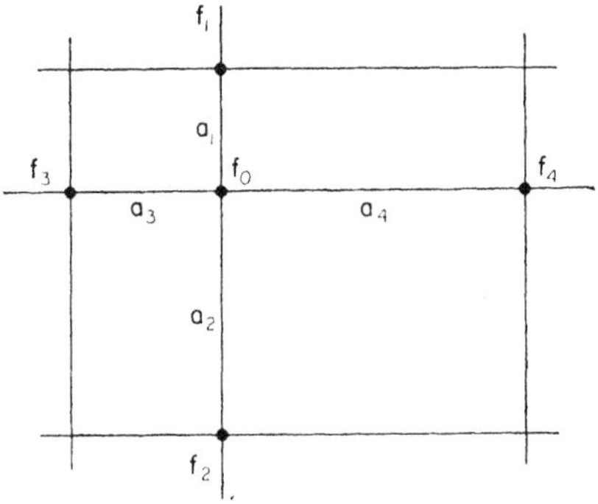

FIG. 4.

Similarly, we may derive the y-derivatives, and hence the value of $\nabla^2 f$ in finite difference form. The third derivative, which is neglected, will give an estimate of the error involved in the solution.

This method is quickly extended to three dimensions or to other systems of coordinates.

LAGRANGE COEFFICIENTS

Suppose we are given the value of a function $f(x)$ at the points x_0, x_1, x_2; then we may write

$$(x) = \frac{(x - x_1)(x - x_2)}{(x_0 - x_1)(x_0 - x_2)} f(x_0) + \frac{(x - x_0)(x - x_2)}{(x_1 - x_0)(x_1 - x_2)} f(x_1) + \frac{(x - x_0)(x - x_1)}{(x_2 - x_0)(x_2 - x_1)} f(x_2).$$

This has fitted a parabola through the three points. If more points had been taken, we could fit an $n-1$th order curve through the n points with the polynomial.

$$f(x) = \sum_{i=1}^{n} \frac{(x-x_1)(x-x_2)\ldots(x-x_n)}{(x_i-x_1)(x_i-x_2)\ldots(x_i-x_n)} f(x_i),$$

where the term $x-x_i$ is omitted from the numerator, and (x_i-x_i) is omitted from the denominator of each ith term. That is, the coefficient of $f(x_i)$ is arranged to equal 1 if $x=x_i$, and zero for any of the other points.

By writing out the polynomial and differentiating, and then substituting the values of x, finite difference forms for the derivative in terms of the points taken arise. The coefficients of each term are called Lagrange coefficients. A short table is given.

Three points:

$$\left.\begin{aligned} 2hf_0' &= 3f_0 + 4\ _1 - f_2 \\ 2hf_1' &= -\ _0 + f_2 \\ 2hf_2' &= f_0 - 4f_1 + 3f_2 \end{aligned}\right\} + \text{the order of 3rd derivative terms.}$$

and $$h^2 f'' = f_0 - 2f_1 + f_2 \text{ at all points.}$$

Four points:

$$\left.\begin{aligned} 6hf_0' &= -11f_0 + 18f_1 - 9f_2 + 2f_3 \\ 6hf_1' &= -2f_0 - 3f_1 + 6f_2 - f_3 \\ 6hf_2' &= f_0 - 6f_1 + 3f_2 + 2f_3 \\ 6hf_3' &= -2f_0 + 9f_1 - 18f_2 + 11f_3 \end{aligned}\right\} + \text{order of 4th derivative terms.}$$

$$\begin{aligned} h^2 f_0'' &= 2f_0 - 5f_1 + 4f_2 - f_3 \\ h^2 f_1'' &= f_0 - 2f_1 + f_2 \\ h^2 f_2'' &= f_1 - 2f_2 + f_3 \\ h^2 f_3'' &= -f_0 + 4f_1 - 5f_2 + 2f_3 \end{aligned}$$

MATRICES

Consider the simultaneous linear equations

$$\begin{array}{l} a_{11}x_1 + a_{12}x_2 + a_{13}x_3 + \ldots + a_{1n}x_n = c_1 \\ a_{21}x_1 + a_{22}x_2 + \ldots \qquad\qquad + a_{2n}x_n = c_2 \\ \cdot \quad \cdot \quad \cdot \quad \cdot \quad \cdot \quad \cdot \quad \cdot \quad \cdot \\ a_{n1}x_1 + a_{n2}x_2 + \ldots \qquad\qquad + a_{nn}x_n = c_n \end{array}$$

It is convenient to consider the quantity, called a matrix, written as

$$\mathbf{A} = (a_{ij}) = \begin{bmatrix} a_{11} & a_{12} \dots & a_{1n} \\ a_{21} & a_{22} \dots & a_{2n} \\ \cdot & \cdot \quad \cdot & \cdot \\ a_{n1} & a_{n2} \dots & a_{nn} \end{bmatrix}$$

The above equation is now written in terms of the matrix **A** and the vectors

$$\mathbf{x} = \begin{bmatrix} x_1 \\ x_2 \\ \cdot \\ \cdot \\ x_n \end{bmatrix} \text{ and } \mathbf{c} = \begin{bmatrix} c_1 \\ c_2 \\ \cdot \\ \cdot \\ c_n \end{bmatrix} \text{ as } \mathbf{A}\,.\,\mathbf{x} = \mathbf{c}$$

or $\sum_j a_{ij} x_j = c_i, i = 1, 2, \dots, n.$

That is,

$$\begin{bmatrix} a_{11} & a_{12} \dots & a_{1n} \\ a_{21} & a_{22} \dots & a_{2n} \\ \cdot & \cdot \quad \cdot & \cdot \\ a_{n1} & a_{n2} \dots & a_{nn} \end{bmatrix} \begin{bmatrix} x_1 \\ x_2 \\ \cdot \\ x_n \end{bmatrix} = \begin{bmatrix} c_1 \\ c_2 \\ \cdot \\ c_n \end{bmatrix}$$

A can therefore be thought of as an operator relating **x** and **c**, or as an array of numbers, or as a kind of numerical name for the set of equations.

If each element is replaced by the element mirroring it in the main diagonal, that is, a_{ij} is replaced by a_{ji}, the resulting matrix is called the transpose **A**′ of the original **A**.

The above defines a simple compact notation for the algebra necessary for dealing with simultaneous linear equations. It can be shown that the sum of two matrices is the matrix composed of the sum of the elements of the two matrices, thus

$$(a_{ij}) + (b_{ij}) = (a_{ij} + b_{ij}).$$

If $\mathbf{A} = (a_{ij})$ and $\mathbf{B} = (b_{ij})$ are two matrices, the product $\mathbf{C} = \mathbf{B}\,.\,\mathbf{A}$ is derived as follows:

The kth member of $\mathbf{A}\,.\,\mathbf{x}$ is $\sum_{j=1}^{n} a_{kj} x_j$.

So the ith member of $\mathbf{B}(\mathbf{Ax})$ is $\sum_{k=1}^{n}\left(b_{ik}\sum_{j=1}^{n} a_{kj}x_j\right)$.

That is, the product $\mathbf{B}.\mathbf{A}=\mathbf{C}=(c_{ij})$ is given by

$$(c_{ij})=\left(\sum_{k=1}^{n} b_{ik}a_{kj}\right).$$

Observe that $\mathbf{BA}\neq\mathbf{AB}$.

The identity matrix is the matrix

$$\mathbf{I}=\begin{vmatrix}1,0,0 & \dots & 0\\ 0,1,0 & \dots & 0\\ . \quad . \quad . & & .\\ 0,0,0 & \dots & 0,1\end{vmatrix}$$

with 1's on the principle diagonal, and zeros everywhere else.

DETERMINANTS

Determinants are numbers which are connected with matrices. Consider the numbers defined by

$$\begin{vmatrix}a_{11} & a_{12}\\ a_{21} & a_{22}\end{vmatrix}=a_{11}a_{22}-a_{12}a_{21},$$

and

$$\begin{vmatrix}a_{11} & a_{12} & a_{13}\\ a_{21} & a_{22} & a_{23}\\ a_{31} & a_{32} & a_{33}\end{vmatrix}$$

$$=a_{11}\begin{vmatrix}a_{22} & a_{23}\\ a_{32} & a_{33}\end{vmatrix}-a_{12}\begin{vmatrix}a_{21} & a_{23}\\ a_{31} & a_{33}\end{vmatrix}+a_{13}\begin{vmatrix}a_{21} & a_{22}\\ a_{31} & a_{32}\end{vmatrix}$$

and so on, where the elements of the first row of the array are taken with alternating sign, multiplying a similar array, derived from the first by leaving out the first row, and the column in which the element occurs.

These numbers are called determinants. To expand the determinant $\det\mathbf{A}=\det(a_{ij})=\|a_{ij}\|$ the following method is used. For the terms in the expansion by the ith row, a_{ij} is multiplied by $(-1)^{i+j}$ and then by the determinant obtained from the first by omitting the ith row and jth column. The terms obtained by setting $j = 1, 2, \ldots, n$ are then summed to give the value of the determinant.

Suppose we have n simultaneous equations defined by $\mathbf{Ax} = \mathbf{c}$. Then, if an inverse $\mathbf{A}^{-1}$ of $\mathbf{A}$ exists such that $\mathbf{A}^{-1}.\mathbf{A} = \mathbf{I}$, and it can be found, we may write

$$\mathbf{A}^{-1}\mathbf{Ax} = \mathbf{A}^{-1}\mathbf{c}$$

or, since $\mathbf{A}^{-1}.\mathbf{A} = \mathbf{I}$, the identity matrix,

$$\mathbf{x} = \mathbf{A}^{-1}.\mathbf{c}.$$

It can be shown that such an inverse matrix exists if and only if the determinant of $\mathbf{A}$ is non-zero. If $\det \mathbf{A} = 0$, the matrix $\mathbf{A}$ is said to be singular, and the solution has to be found by different means, depending on the nature of $\mathbf{A}$.

If we want to solve the equation $\mathbf{Ax} = 0$, clearly $\mathbf{x} = 0$ is a trivial solution. The condition that a non-zero solution exists is that $\det \mathbf{A} = 0$.

EIGENVALUES AND EIGENFUNCTIONS

Consider the relation $\mathbf{A}.\mathbf{x} = \mathbf{y}$ connecting the two vectors $\mathbf{x}$ and $\mathbf{y}$. If any vectors exist such that $\mathbf{A}.\mathbf{x} = \lambda\mathbf{x}$, where λ is a scalar we may write

$$(\mathbf{A} - \lambda\mathbf{I})\mathbf{x} = 0.$$

For non-trivial solutions, $\det(\mathbf{A} - \lambda\mathbf{I}) = 0$.

This is a polynomial of degree n, which may be solved for its n roots. These roots λ_i are called the eigenvalues or the characteristic roots of the matrix $\mathbf{A}$, and the corresponding vectors $\mathbf{x}_i$ obtained by solving $(\mathbf{A} - \lambda\mathbf{I})\mathbf{x}_i = 0$ are called the eigenvectors. They have the property that any vector $\mathbf{x}$ can be written as a linear combination of the eigenvectors, $\mathbf{x} = \alpha_1 \mathbf{x}_1 + \alpha_2 \mathbf{x}_2 + \ldots + \alpha_n \mathbf{x}_n$, and that no eigenvector can be written as a linear sum of the other eigenvectors. This is summed up mathematically by saying the eigenvectors span the space of vectors considered, and that they are orthogonal.

A mention was made of eigenfunctions earlier. These two properties of spanning and orthogonality apply to any set of eigenfunctions, in a modified form. The best-known set of eigenfunctions is that used in expanding in Fourier series, $\sin n\theta$ and

$\cos n\theta$, where n goes from 1 to infinity. It can be proved that this set of functions span the set of all well-behaved functions with period 2π, and they satisfy the orthogonality relations

$$\int_0^{2\pi} \sin n\theta \sin m\theta \, d\theta = 0 \quad \text{if } n \neq m$$
$$= \pi \quad \text{if } n = m.$$
$$\int_0^{2\pi} \sin n\theta \cos m\theta \, d\theta = 0$$
$$\int_0^{2\pi} \cos n\theta \cos m\theta \, d\theta = 0 \quad \text{if } n \neq m$$
$$= \pi \quad \text{if } n = m.$$

These relations enable us to expand a periodic function $f(\theta)$ in the form

$$\tfrac{1}{2}A_0 + \sum_1^\infty A_n \cos n\theta + \sum_1^\infty B_n \sin n\theta,$$

where $A_n = \dfrac{1}{\pi}\displaystyle\int_0^{2\pi} f(\theta)\cos n\theta \, d\theta, \quad B_n = \dfrac{1}{\pi}\displaystyle\int_0^{2\pi} f(\theta)\sin n\theta \, d\theta.$

REFERENCES

Journée Int. de Calcul Analogique, 1955 Conference at Brussels, *Proceedings*.

R. C. HANSON *et al.*, On computing radiation integrals, *Assoc. for Computing Machinery*, Vol. 2, No. 2 (Feb. 1959).

MILNE, *Numerical Calculus*.

JEFFRIES and JEFFRIES, *Methods of Mathematical Physics*, Cambridge University Press.

D. G. RUTHERFORD, *Vector Methods*, Oliver & Boyd.

PROF. C. A. COULSON, *Electricity*, Oliver & Boyd.

PROF. A. C. AITKEN, *Determinants and Matrices*, Oliver & Boyd.

E. G. PHILLIPS, *Functions of a Complex Variable*, Oliver & Boyd.

H. T. H. PIAGGIO, *Differential Equations*, G. Bell & Son.

Modern Computing Methods, National Physical Laboratory.

N. W. MCLACHLAN, *Bessel Functions for Engineers*, Oxford University Press.

J. C. JAEGER, *An Introduction to the Laplace Transformation*, Methuen.

B. J. STARKEY, *Lapace Transforms for Electrical Engineers*, Iliffe.

B. HAGUE, *An Introduction to Vector Analysis*, Methuen.

P. H. HAMMOND, *Feedback Theory and its Applications*, E.V.P.

G. J. THALER and M. P. PASTEL, *Analysis Design of Nonlinear Feedback Control Systems*, McGraw-Hill.

chapter 2

ELECTRONIC ANALOGUE EQUIPMENT

THE AMPLIFIER

THE D.C. AMPLIFIER

The d.c. amplifier is the basic building block of the electronic analogue computer. It can be made to sum, multiply by constants, and integrate voltages, and to drive such things as multipliers, voltmeters and recorders in associated equipment. Much design effort has gone into the perfection of low noise, stable, drift-free amplifiers with band widths of up to 100 kc/s. Such an amplifier is a precision piece of equipment, very reliable, and useful in many fields other than analogue computation.

Consider Fig. 5a; the round-backed triangle is the conventional symbol for a d.c. amplifier. We assume for the moment that it draws no current at its input A, which will in practice be the grid of its first valve, that it has an effectively infinite negative gain, and that it does not drift. If a small voltage V_g exists at A, a "very large" negative voltage will be produced at the output B, and a current will flow along AR_0B until V_g becomes sensibly zero. This input point A is, therefore, called the virtual earth; for a practical computing amplifier with a gain of perhaps one million, it will never be more than about 0·1 mV from earth, which will, therefore, give an output of 100 V.

The amplifier in Fig. 5a is fed by a voltage V_1 through a

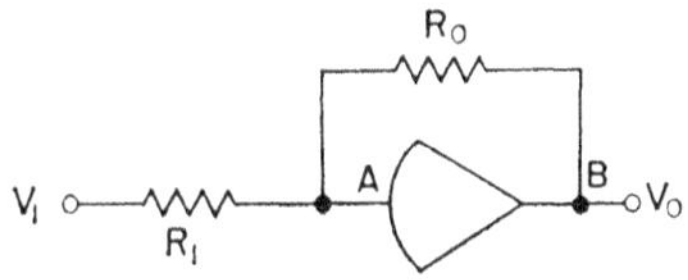

(a)

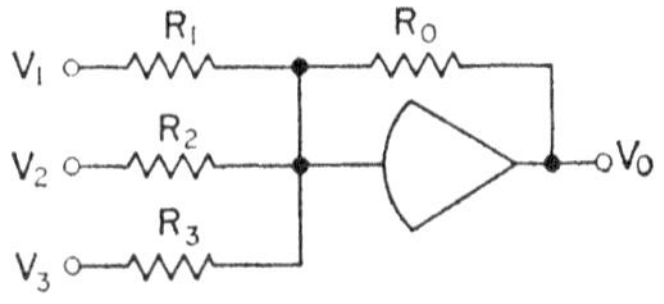

$$V_0 = -\left(\frac{R_0}{R_1}V_1 + \frac{R_0}{R_2}V_2 + \frac{R_0}{R_3}V_3\right)$$

or $$V_0 = -(\gamma_1 V_1 + \gamma_2 V_2 + \gamma_3 V_3)$$

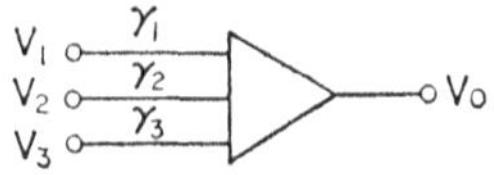

$$V_0 = -(\gamma_1 V_1 + \gamma_2 V_2 + \gamma_3 V_3)$$

(b)

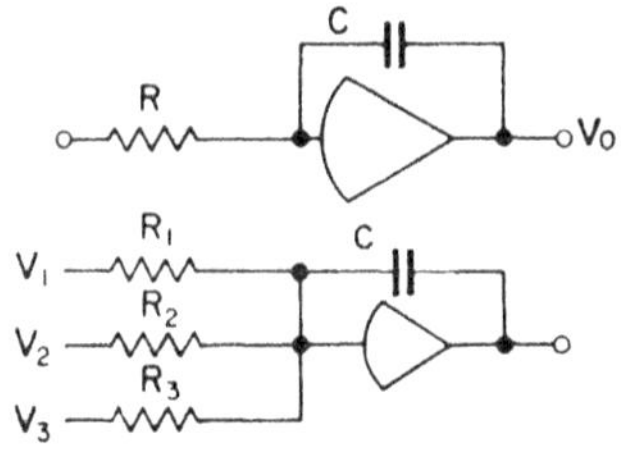

$$V_0 = -\frac{1}{pC}\left(\frac{V_1}{R_1} + \frac{V_2}{R_2} + \frac{V_3}{R_3}\right)$$

or $$V_0 = -\frac{1}{p}(\gamma_1 V_1 + \gamma_2 V_2 + \gamma_3 V_3)$$

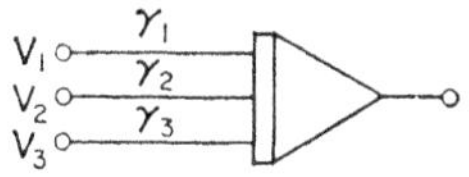

$$V_0 = -\frac{1}{p}(\gamma_1 V_1 + \gamma_2 V_2 + \gamma_3 V_3)$$

(c)

FIG. 5.

resistor R_1, and has a feedback resistor R_0 from the output voltage V_0. Writing down a current balance for the virtual earth A, we have from Kirchhoff's law,

$$\frac{V_1}{R_1}+\frac{V_0}{R_0}=0.$$

Therefore $$-V_0=\frac{R_0}{R_1}V_1.$$

Clearly, if voltages $V_1, V_2, V_3, \ldots,$ feed the virtual earth through input resistors $R_1, R_2, R_3, \ldots,$ we have (Fig. 5b)

$$-V_0=R_0\left\{\frac{V_1}{R_1}+\frac{V_2}{R_2}+\frac{V_3}{R_3}+\ldots\right\}$$

or $$-V_0=R_0\sum_i\frac{V_i}{R_i}$$

or $$-V_0=\sum_i\gamma_i V_i,$$

where $\gamma_i=R_0/R_i$ with the dimension of a pure number, is called the gain of the ith input, and the summations are taken over all inputs.

An amplifier connected with resistors in this manner is therefore seen to act as a summer of the input voltages, with a sign inversion on the sum. The conventional symbol for a summing amplifier with gains $\gamma_1, \gamma_2, \gamma_3$ is shown in Fig. 5b.

If we consider the general case of an amplifier with input impedances Z_i and a feedback impedance of Z_0, the current balance still holds, exactly as in the case of the summing amplifier. It can be seen that in this case we obtain

$$-V_0=Z_0\sum_i\frac{V_i}{Z_i},$$

where the summation is over all inputs.

If the input impedances are resistances R_i, and the feedback impedance is a pure capacity C, this equation becomes that of an integrator, thus:

$$-V_0=\frac{1}{pC}\cdot\sum_i\frac{V_i}{R_i}=\sum_i\frac{V_i}{pCR_i}$$

or $$V_0 = \sum_i \gamma_i \frac{V_i}{p},$$

where $$\gamma_i = \frac{1}{R_i C}.$$

Since the transfer function between V_0 and the ith input is

$$\frac{1}{pCR_i} = \frac{\gamma_i}{p},$$

which is that of an integrator, such an amplifier gives the sum of the integrals of the input voltages with respect to time. The constants γ_i, with dimensions of number divided by time, are called the gains of the integrator.

Writing down the differential equations for one input of an integrating amplifier, we have, alternatively (Fig. 5c)

$$\frac{V_1}{R_1} + C\frac{dV_0}{dt} = 0.$$

Therefore $$-V_0 = \frac{1}{R_1 C}\int_{t_0}^{t_1} V_1 \, dt,$$

where t_0 and t_1 are the starting time and the time of observation.

The conventional symbol for an integrating amplifier with integrator gains γ_1, γ_2, γ_3, is shown in Fig. 5c.

By putting a capacitor in the input and a resistor as the feedback component of an amplifier, a transfer function of the form pCR is obtained. This corresponds to a pure differentiator. However, a differentiator accentuates any noise that may be present in the circuit, since its gain increases with frequency. Crudely, the differential of a spikey waveform is a much more spikey waveform. There is always some noise present at 50 c/s for a reason to be explained below, and the gain of a differentiator at 50 c/s is 100 times that at 0·5 c/s, which is a high frequency for an analogue problem, except in special applications.

For this reason, pure differentiators are almost never successful in analogue circuits. All equations are transformed into terms of integration as the operation for solution of the differential equations for the system variables. Occasionally a differentiation is unavoidable, and it must be approximated; this will be dealt with in a later chapter.

BASIC RELAY CONTROLS OF THE AMPLIFIER

We have shown how to use the amplifier for summing or integrating. In order to control the amplifier operation, relays are used to switch inputs to the amplifier. Figure 6 shows a typical arrangement. The common point of the input resistors,

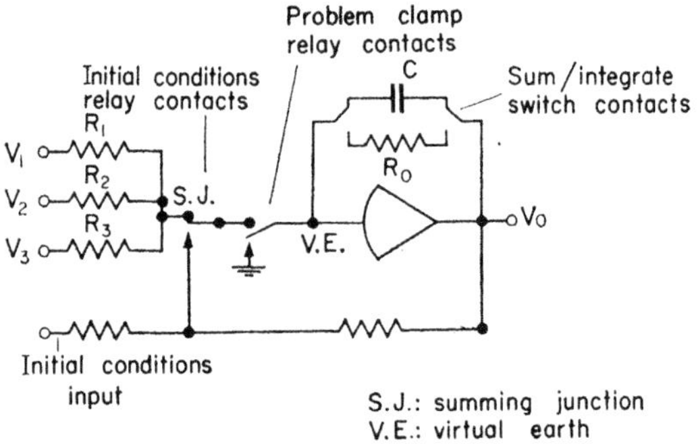

FIG. 6.

called the summing junction, is connected to the virtual earth via two relay contacts. For the amplifier as shown, the problem clamp relay connects the virtual earth direct to earth. The output of the amplifier, therefore, remains constant, or clamped, at the voltage across its integrating feedback capacitor. If the relay is changed over to make contact to the summing junction, integration starts. If the initial condition relay is changed over also, the amplifier gives as output a constant voltage determined by V_i, the initial conditions input. Releasing this relay then gives integration from this voltage as an initial condition.

Two further contacts enable the amplifier to be switched to give a capacitor or a resistor as feedback component, for its use as an integrator or as a summer. The operation of these contacts is self-explanatory.

By means of more complex relay circuitry, computers are arranged to have various conditions or modes of operation. In the "operate" or "compute" condition, summing junction and virtual earths are connected for all amplifiers. In the "hold", "clamp" or "freeze" conditions, summers are in the operate condition, but the virtual earths of all integrators are earthed. In the "gain check" or "rate test" condition, all amplifiers are switched to the summing conditions, and the gains of integrators as well as summers can then be adjusted. "Set initial conditions" switches the integrators to charge their feedback capacitors to the required initial voltage, using an additional initial condition input. Finally, "standby" or "zero set" switches all amplifiers to the summing conditions, with only one input to each; those inputs are earthed, and arranged to give a high gain of perhaps 1000. The amplifiers can then be trimmed to ensure that they give zero volts output for zero volts input.

EXERCISES

1. Modify the circuit of Fig. 6 to give a "zero set" facility.
2. The facilities of "zero set", "operate", "gain check", "hold" and "set initial conditions" are required for a computer with ten amplifiers. By operating switches, five out of the ten can be switched to be either summers or integrators, whilst the remainder are always summers. Devise relay circuits to give the required facilities, which are to be initiated by the operator closing a switch for the required condition.

ERRORS OF THE SIMPLE AMPLIFIER

The primary cause of error in d.c. computing amplifiers is drift. The circuit of a simple form of amplifier is given in Fig. 7. Suppose the input voltage V_1 is made zero and the potentiometer R_v is

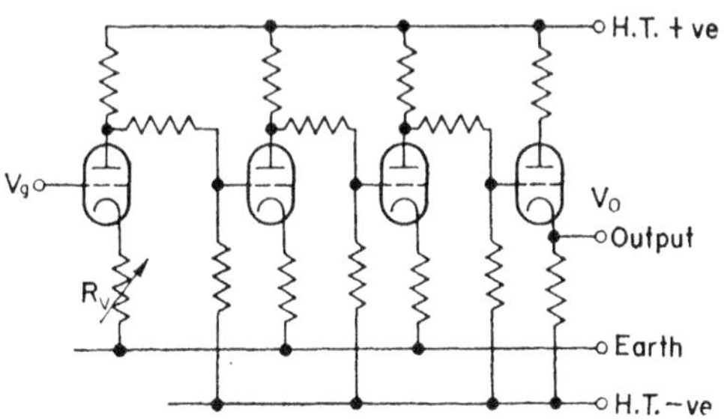

FIG. 7.

adjusted until the output V_0 is made zero. If the amplifier is now left, the output will drift from zero because of such things as slow changes in the properties (the g_m) of the valves, alterations of the H.T. supply, or mains voltage drift resulting in heater current changes, temperature changes and so forth. The amplifier is particularly sensitive to changes affecting the first valve, a small alteration here affecting all subsequent voltage levels. Good design can reduce the drift of such a straight d.c. amplifier to values of $100\mu V$ per day; the techniques used are ready aged valves, balanced stages of "long-tailed pair" type, very good voltage regulation and temperature control. But the invention and perfection of the chopper stabilization method by E. A. Goldberg in the 1950's vastly improved the performance of the d.c. amplifier, and made it into a reliable and accurate piece of equipment.

THE CHOPPER STABILIZED AMPLIFIER

An a.c. amplifier is basically free from drift, since it does not pass d.c. However, it can be used to give a drift-free d.c. amplifier in the following way. By using a chopper relay, the input of an a.c. amplifier is fed alternately from an input voltage V_1, and from earth, at a rate of, say, 50 c/s. This a.c.signal is amplified, and then rectified and smoothed using a simple RC filter, and a second contact of the chopper relay to act as the rectifying element.

The frequency response of this amplifier is limited by the chopper frequency and the smoothing filter, so that it is not, by itself, a practical amplifier. It must, therefore, be combined with a d.c. amplifier to increase its band width.

Consider the combination shown in Fig. 8. The virtual earth of a d.c. amplifier is fed to a chopper relay, which feeds an a.c. amplifier with V_g and zero alternately. The waveforms at A, B and C illustrate the actions of the a.c. amplifier and the chopper relay. to give voltages V_A, V_B and V_C.

The d.c. amplifier has a balanced differential input, with, in fact, two separate inputs. The output is given by the gain times the difference between the two input voltages. The d.c. amplifier is connected so that the smoothed output of the a.c. amplifier V_c

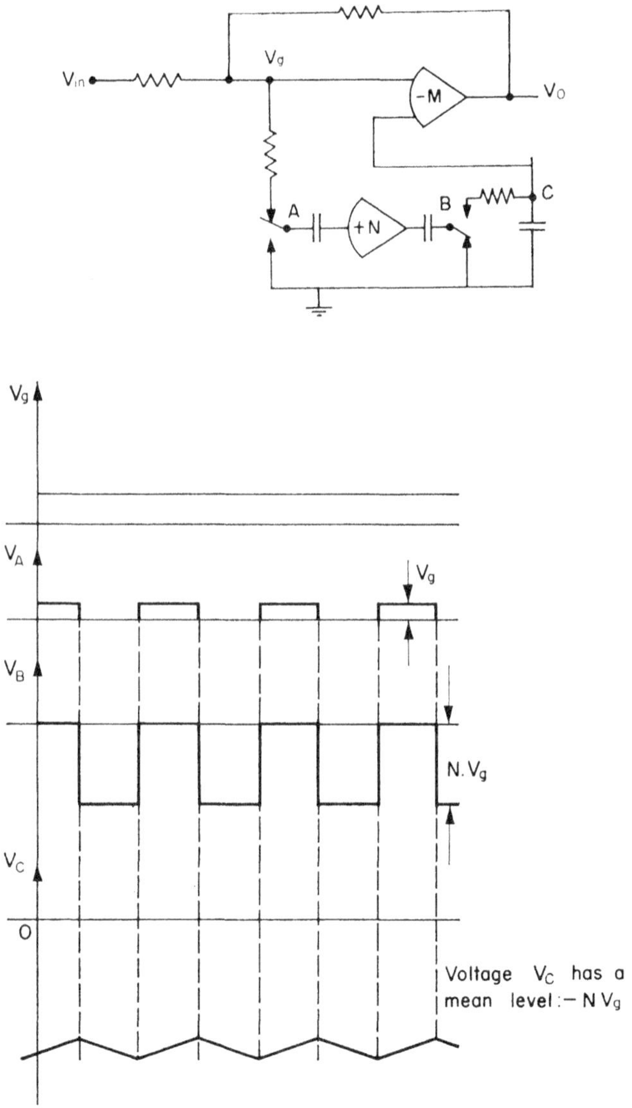

Vin
Vg
-M
Vo
A
+N
B
C
Vg
VA
Vg
VB
N. Vg
VC
O
Voltage VC has a mean level: − N Vg

FIG. 8.

now feeds its differential input stage, i.e. the amplifier now amplifies $(V_g - V_c)$ by $-M$, giving output $V_c = -M(V_g - V_c)$. Now $V_c = -NV_g$, so the final output is

$$V_c = -M(N+1)V_g.$$

This gives a much greater gain, but it does more than that, for the drift is present in the d.c. stage only. If the total drift voltage of the d.c. stage is V_d, referred back to the input as if it occurred there, we have

$$V_0 = -M\{V_g + V_d - (-NV_g)\}.$$

Therefore $$V_0 = -M(N+1)V_g - MV_d.$$

Thus the gain for the drift term is reduced by a factor $N+1$ compared to the total gain from the combined a.c. and d.c. amplifying stage.

In practice typical values are $N = 2000$ and $M = 30{,}000$, giving a total d.c. gain 6×10^7, the ratio of drift to signal being improved by a factor of 2000 over that of the d.c. stage alone. The circuit of a modern d.c. amplifier is given in Fig. 9. The input stage of the d.c. part is a long-tailed pair, the virtual earth being taken direct to a valve taking a very low grid current, and the smoothed a.c. amplifier output going to the grid of the other valve of the pair. The output stage is a cathode follower, which can give up to 10 mA without distortion in the range +100 to −100 V at an output impedance of 1000 Ω or so. This impedance is reduced to the order of a milliohm by the high negative feedback introduced by the external feedback component.

The presence of the synchronous chopper relay circuit introduces noise at the chopper frequency. This is normally the mains frequency, the chopper being driven from the heater supply, although 400 c/s choppers are used occasionally.

ERROR ANALYSIS

Consider an amplifier with voltage V_g at its input grid and gain $-M$, drawing no grid current.

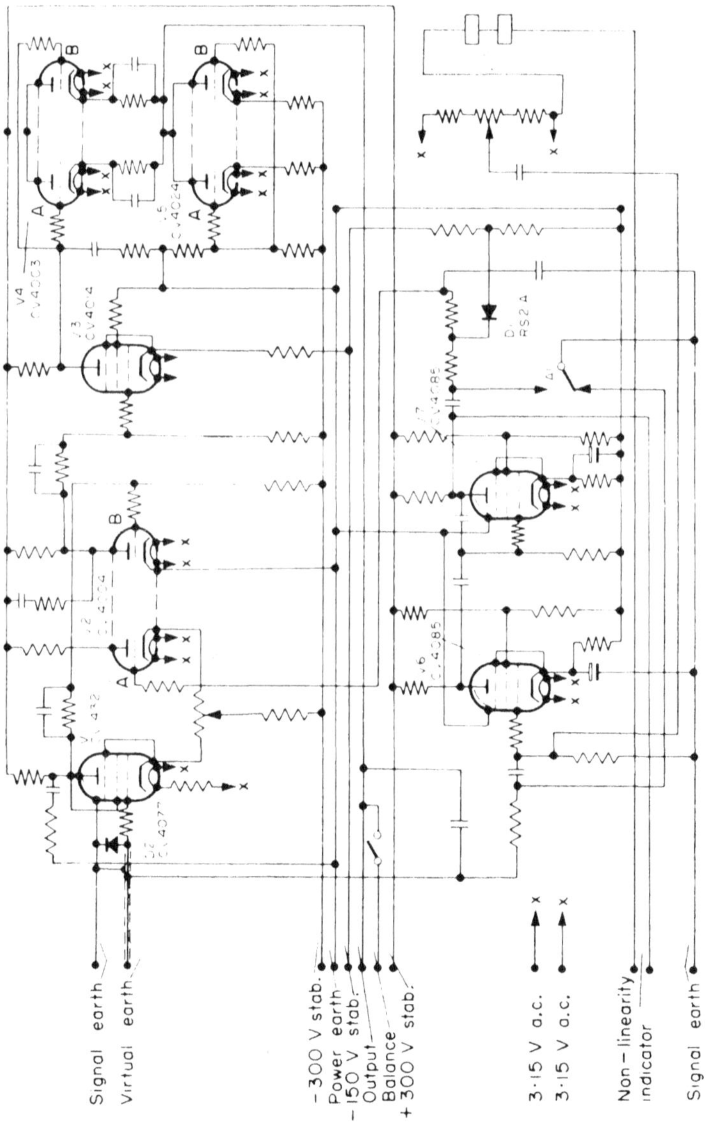
Signal earth
Virtual earth
- 300 V stab.
Power earth
- 150 V stab.
Output
Balance
+ 300 V stab.
3·15 V a.c.
3·15 V a.c.
Non-linearity indicator
Signal earth
A
B
A
B
A
B

FIG. 9.

Then the current balance gives

$$\sum_{i=0}^{n} \frac{V_i - V_g}{Z_i} = 0, \quad \text{where } Z_0 \text{ is the feedback impedance.}$$

Also, $$V_0 = MV_g.$$

Therefore $$\sum_{i=1}^{n} \frac{V_i}{Z_i} = -\frac{V_0}{Z_0} - \frac{V_0}{M}\sum_{i=0}^{n} \frac{1}{Z_i}.$$

For an amplifier with five 1 MΩ inputs, two 100 kΩ inputs, two 500 kΩ inputs and a 1 MΩ feedback resistor, which will give gains of 1, 10 and 2, the quantity

$$\sum_{1}^{10} \frac{1}{Z_i} = 30\,(\text{M}\Omega)^{-1}.$$

For an amplifier gain of $M = 10^7$ times, the error term

$$\frac{V_0}{M} \cdot \sum \frac{1}{Z_i}$$

in the above equation for V_0/Z_0 gives an error current of $3 \times 10^{-6} V_0$ μA with the resistor values given. Now the output V_t of a perfect amplifier would be given by

$$-\frac{V_t}{Z_0} = \sum_{1}^{n} \frac{V_i}{Z_i},$$

and so $$V_t = V_0 + \frac{V_0}{M} \cdot Z_0 \sum_{0}^{n} \frac{1}{Z_i}.$$

Therefore the actual voltage for a 1 MΩ feedback resistor is related to the ideal voltage V_t by

$$V_t = V_0(1 + 3 \times 10^{-6}),$$

which is an error of 3×10^{-4} per cent of the output voltage.

If there is drift in the amplifier, it can be referred back to the input, and thought of as due to a voltage V_d at the grid.

Then $V_0 = -M(V_g + V_d)$, which leads to

$$\sum_{i=1}^{n} \frac{V_i}{Z_i} = -\frac{V_0}{Z_0} - \frac{V_0}{M}\sum_{i=0}^{n} \frac{1}{Z_i} - V_d \sum_{i=0}^{n} \frac{1}{Z_i}.$$

Using the resistor values given above, the error due to drift is thus $30V_d$ Volts. A value of $100\,\mu$V for V_d is fairly realistic, giving an error of 3×10^{-3} V, or 3×10^{-3} per cent of a 100 V output.

EXERCISE

Show that the error due to a grid current of $4 \times 10^{-5}\,\mu$A is negligible compared with the drift error, for both a summer and an integrator with feedback capacity of $1\,\mu$F.

In addition to the above causes of error, which directly involve the amplifier, the wiring forming the interconnections between amplifiers will have a leakage resistance of the order of $10^5\,\mathrm{M}\Omega$, giving leakage currents of $10^{-3}\,\mu$A. This means that the leakage of an integrator will be 10^{-5} V/sec, or 1 per cent in 1000 sec.

EXERCISE

The output impedance of a cathode follower output stage of a computing amplifier is $500\,\Omega$. If the amplifier gain is 10^7, show that the output impedance with a $1\,\mathrm{M}\Omega$ feedback is about $2{\cdot}5 \times 10^{-4}\,\Omega$.

ERRORS OF SUMMERS AND INTEGRATORS

The value of the gain of the d.c. stage of an amplifier falls off with increasing frequency; to a fair approximation, we may write

$$M(p) = -\frac{M}{1+pT_1}, \quad N(p) = -\frac{N}{1+pT_2},$$

where $T_1 = 3{\cdot}3 \times 10^{-3}$ sec, $T_2 = 20$ sec.

Thus the total gain is

$$M_t(p) = \frac{M(1+N)\{1+(pT_2/1+N)\}}{(1+pT_1)(1+pT_2)} = M_t\frac{(1+pT_3)}{(1+pT_1)(1+pT_2)},$$

where $T_3 = 0\;01$ sec.

Referring to the formula for finite gain error, the error is the inverse of the gain, so we obtain the error curve shown in Fig. 10a using the asymptote approximation.

EXERCISE

Show that the integrator error curve is as shown in Fig. 10b for an integrator with input resistors as given in the text, and a feedback capacitor of $1\,\mu$F.

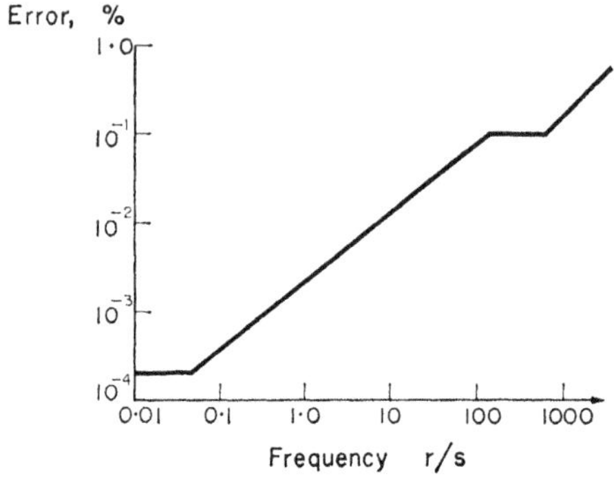

FIG. 10a. Summer errors.

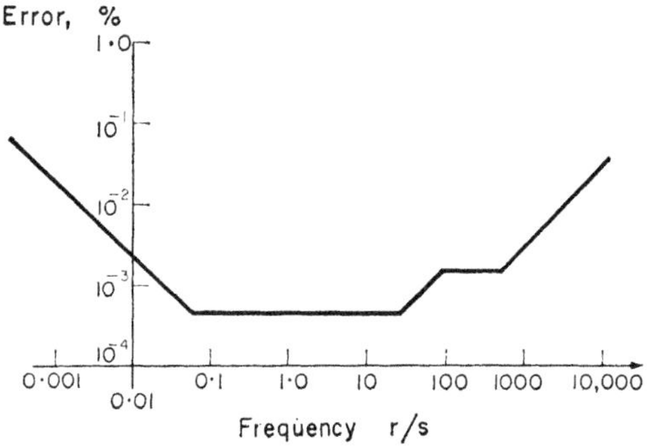

FIG. 10b. Integrator errors.

Thus, at 100 r/s, or 16 c/s, a summer will have less than 0·1 per cent error, and the integrator error will be less than 0·1 per cent from 5×10^{-4} r/s to 10^4 r/s.

The noise introduced by the chopper will be of the order of 10 mV for a summer, 2 mV for an integrator.

MULTIPLIERS

THE COEFFICIENT POTENTIOMETERS

The simplest form of multiplication of a voltage is that given by tapping off a fraction of it using a potentiometer. If, in Fig. 11, $R_0 = R_1$, and the potentiometer slider was moved from one end of its track to the other, the output V_0 from the amplifier would vary from nothing to $-V_1$, where V_1 is the voltage above the potentiometer. Thus, if we wished V_0 to be given by $V_0 = -\gamma V_1$, by putting a reference voltage of 100 V as V_1, and reading V_0, the slider could be moved until the desired fraction was found.

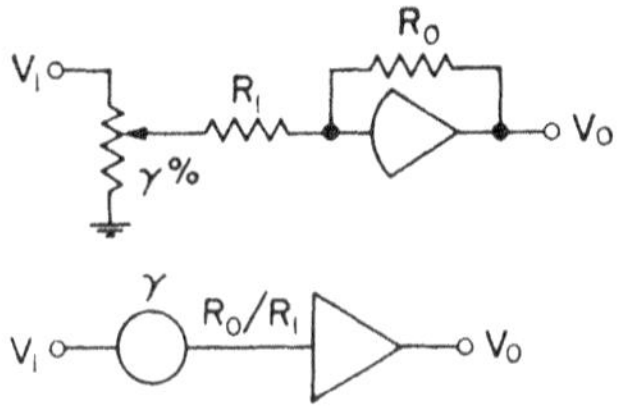

FIG. 11.

Of course, the resistor R_1 to the amplifier virtual earth will load the potentiometer, and the loading varies with slider position, so that the value of V_0 is not a linear function of position. But this is of no importance, since we are not interested in the position of the slider, only in the behaviour of the unit as a whole.

The conventional symbol for a potentiometer is shown in Fig. 11. It is standard practice to arrange the gains of an amplifier to be exact numbers such as 10, 1 or 0·1. To set the gain of an amplifier to some other value, a coefficient potentiometer is used, and set up in the loaded condition either using a high input impedance voltmeter on the slider or by reading the amplifier output, with a standard voltage at input.

THE SERVO-MULTIPLIER

If the slider of a potentiometer is driven mechanically to a position proportional to a voltage V_x and a voltage V_y is put across the potentiometer, its slider will be at a voltage proportional to the product $V_x V_y$. This may be done by a position servo, as shown in Fig. 12. The amplifier takes V_x and compares it with V_s from a position potentiometer which has the reference voltage, usually 100 V, across it. The output of the amplifier feeds a drive amplifier, which drives a motor whose shaft carries the potentiometer sliders round the potentiometer tracks. The motor will turn until it has reduced the amplifier input $V_x - V_s$ to zero. The potentiometers are then positioned at a proportion

$V_x/100$ of their travel, and the sliders of the ganged potentiometers will, when suitably loaded, reproduce this proportion of the voltage across them. Thus we obtain the product in the form $V_x V_y/100$ from an output amplifier whose input resistor matches the inputs to the multiplier drive amplifier. Conventional symbols are shown in the figure.

In practice the drive amplifier will have a velocity feedback input as well, to improve the response, and various shaping networks may be added. A good multiplier can be made to have a static accuracy of about 0·5 per cent, and an amplitude response which is 3dB down at 400 r/s. However, the phase response is

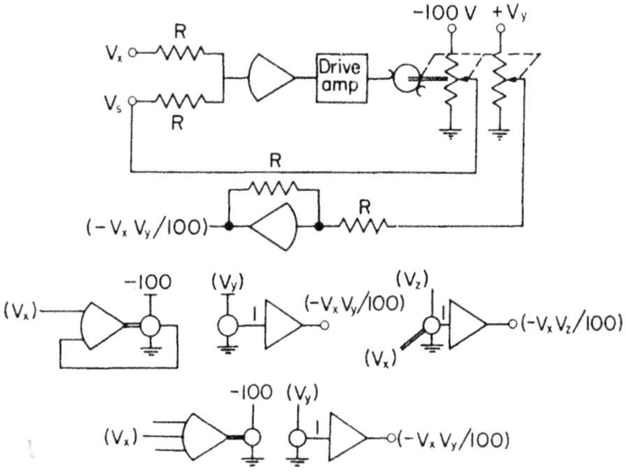

FIG. 12.

2–3° in error at 10 r/s, which gives the effective limit, since a 2° phase lag in the tracking of the multiplier will give a large error in the product. This may be seen by considering a 100V peak sine-wave drive, with ± 100V on the first multiplying slave potentiometer. When the input sine wave reaches 0V, the slider will be at a proportion of $\sin 2° \approx 0{\cdot}03$ of the travel, giving an output of 3V, or 3 per cent error.

The step response of a multiplier of this type gives typical rise times of the order of 20msec; following a step input to the drive, the sliders would have attained 90 per cent of their final movement in about this time.

The advantages of the servo-multiplier compared with the many other methods which have been devised for multiplication are reasonable cost, reliability and the ease with which it can multiply several variables by the same input quantity, merely by ganging several potentiometers to the motor shaft. They are mechanically simple, and easily maintained, and, with proper protection circuits to prevent continuous rotation with the consequent wear on the potentiometer tracks, are almost foolproof. Their disadvantages are that they are slow, of limited accuracy, and they introduce a certain amount of noise.

Figure 16a shows circuits for four quadrant multiplication, division and square rooting.

EXERCISE

In the divide and square-root circuits given in Fig. 16 what are the input voltages to the amplifiers from the slave potentiometers? Devise circuits to give cube roots and other non-integral powers.

CROSSED-FIELD MULTIPLIERS

Special cathode ray tubes having perpendicular electric and magnetic field are produced for multiplication. An electron gun produces a beam of electrons with about 1·5 kV as the accelerating field. This beam is deflected in the X-direction by the input V_x on the X plates; they therefore have an X-component of velocity, which interacts with an axial magnetic field proportional to V_y. This is produced by a coil wound round the tube carrying a current proportional to V_y. The interaction gives a deflection in the Y-direction. This deflection is then reduced to nothing by a voltage applied to the Y plates, and it is easily shown that this voltage is proportional to the product $V_x V_y$. The detection of the deflection can be done by masking half the screen of a phosphorescent cathode ray tube and amplifying the output of a photocell to give the Y-voltage in such a sense as to tend to bring the beam of electrons just to the edge of the mask. Special electrodes within the tube may also be used. Generally a pair of plates, one above the X-axis and one below, are fitted within the tube, and an electronic servo brings the beam on to the axis by balancing the currents in the two plates. This method has the advantage that it does not depend on the behaviour of the phosphor.

Such multipliers are fast, their accuracy depending on the design of the electrodes (for low secondary electron current from the detecting plates to the Y plates, and correct axial field), and the amplifiers driving the coil and the Y plates. A typical performance is 1° phase shift at 2 kc/s, 1 per cent accuracy in that range. They need careful screening from stray fields, and careful setting up. They have been used in computers and response analysers to give the reference and quadrature components of a system output when driven by a sine-wave input.

HALL EFFECT MULTIPLIERS

If a rectangular semiconductor wafer in the $X-Y$ plane has steady current I flowing in its X-direction, and a magnetic field B exists perpendicular to the plane of the wafer, a small voltage will appear across it in the Y-direction. This voltage is called the Hall voltage, after its discoverer, and is proportional to the product IB. More exactly,

$$V = \frac{R_h IB \times 10^{-8}}{t},$$

where I is in amps, B is gauss, V in volts, t is the wafer thickness in cm, and R_h is the Hall coefficient in cm^3/coulomb. Using transistor amplifiers, several multipliers have been built of very similar performance to the crossed field cathode ray tube multiplier already described.

QUARTER-SQUARE MULTIPLIERS

The quarter-square multiplier depends on the relation

$$xy = \tfrac{1}{4}\{(x+y)^2 - (x-y)^2\}$$

for its operation.

If the squares of the sum and the difference of the inputs V_x and V_y can be formed, a subtraction will give the product. Passive circuits can be constructed to perform the squaring from diode circuits which will be described later. It is then a simple matter to obtain the product voltage.

These multipliers can be made fast and accurate. For small inputs, the error can be large unless the circuit is modified somewhat, due to the flatness of the square curve near the origin. This is overcome by the following method in the Reeves multiplier. We have

$$4xy = [3(x+y)+(x+y)^2]-[3(x-y)+(x-y)^2]-6y.$$

The two quantities in the square brackets are produced by diode networks and the final addition gives the product. The advantage is that the two curves of $3(x+y)+(x+y)^2$ and $3(x-y)+(x-y)^2$ cut the origin at an angle, thus greatly improving the performance at small signal levels.

Triodes can be made to perform the squaring by using the fact that the anode current is approximately proportional to the anode voltage squared. This and similar methods are often used in repetitive computers.

TIME-DIVISION MULTIPLIERS

If a repetitive pulse has its width made proportional to one variable and its height proportional to the other, its area is a measure of the product. If this is done fast enough, and modern techniques make pulse repetition rates of 20 kc/s possible, an extremely accurate, fast multiplier results. The pulse may be triangular, or the mark-space ratio of a square pulse can be varied; the output is then smoothed in an *RC* circuit. There are various subtleties of the electronics in the practical design, which are beyond the scope of the present book and will not be described here.

The electronics of these multipliers must be watched, as they readily loose their adjustment. But they are widely used and are capable of accuracies of the order of 0 05 per cent full scale from 0 to 10 kc/s.

If V_x is made to vary the pulse duration, this pulse train can be sent to several modulators to give pulse heights proportional to V_{y_1}, $V_{y_2}, \ldots$, so that several products by V_x can be formed as in the servo-potentiometer multiplication, to give products $V_xV_{y_1}$, $v_xv_{y_2}$, etc.

DIODE FUNCTION GENERATORS AND LIMITERS

The gain of an amplifier can be varied as a function of the input voltage by arranging for diodes to switch in resistors in the feedback or input network. Consider the curve in Fig. 13a. It can be simulated crudely by the two straight lines shown, the transition or "break" from one to the other being at 50 V. The amplifier in Fig. 13b can be arranged to give an output voltage

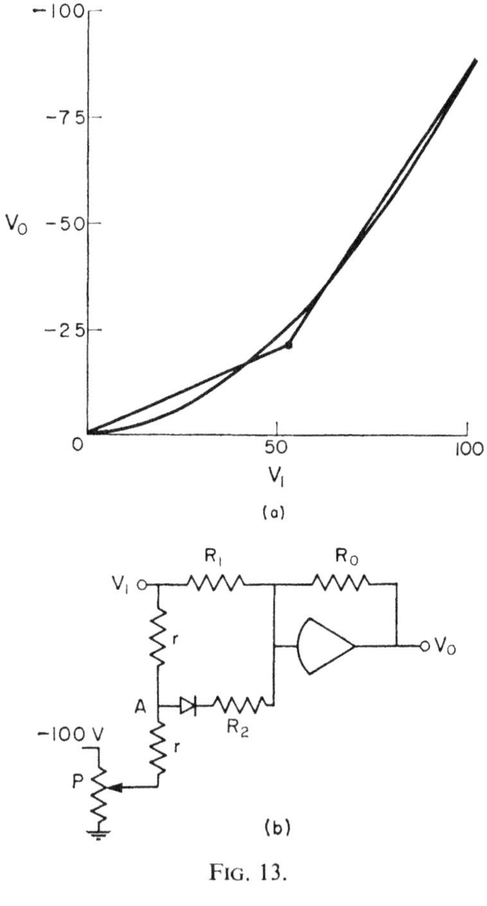

FIG. 13.

defined by the two straight lines. If the slider of the potentiometer P is at -50V and V_1 is just below 50V, at 49V, the point A is below zero, and the amplifier gain is R_0/R_1. If V_1 is at 51V, A is above zero and the diode conducts. The gain them becomes $R_0/R_1+R_0/R_2$, if r is small compared with R_2.

Thus, to simulate our function

$$\frac{R_0}{R_1}=\frac{25}{50}$$

$$\frac{R_0}{R_1}+\frac{R_0}{R_2}=\frac{75}{50}.$$

So, putting $R_0 = 1\,\text{M}\Omega$, we find $R_1 = 2\,\text{M}\Omega$, $R_2 = 1\,\text{M}\Omega$ gives the required approximation to the curve.

Clearly, the number of diode networks of this type can be increased. About ten is usually sufficient to give a simulation of 0·25 per cent accuracy for a wide range of functions.

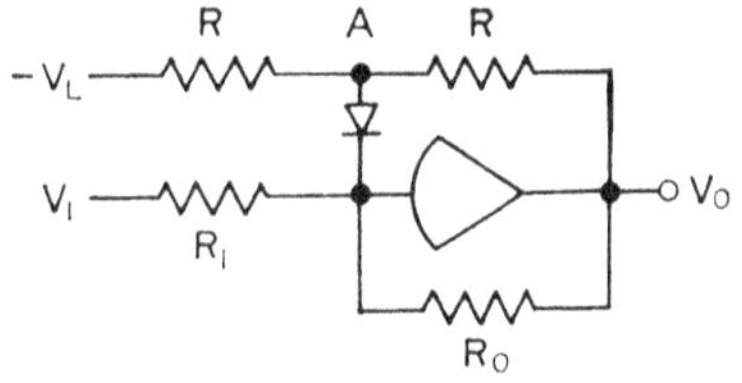

FIG. 14.

On a typical function generator of this type, both signs of the input voltage and $+100$V and -100V will appear on specified sockets or sets of tags. Also, sets of potentiometers for adjusting the break points where each straight line segment takes over from its predecessor, and potentiometers to adjust the gains to give the right slopes to those segments will be fitted.

The limiter is a special form of function generator. An amplifier with a diode arranged as in Fig. 14 will have a gain of R_0/R_1 so long as the output remains below V_L. If the output increases further, the point A will then rise above zero, and the diode will conduct. The output then remains at V_L provided R is small compared to R_0.

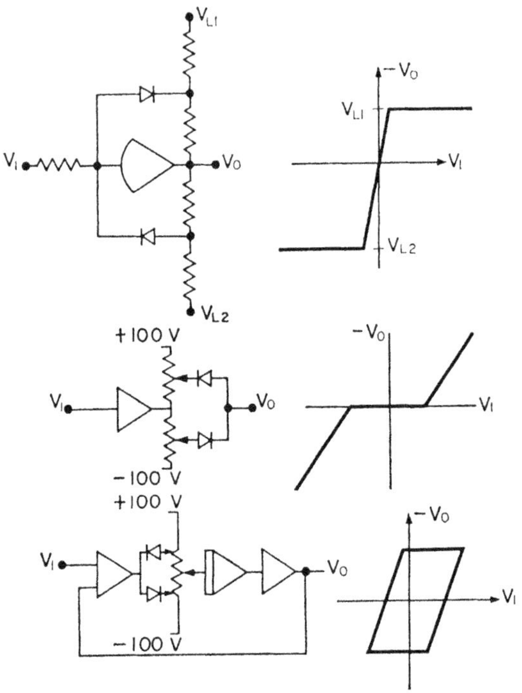

FIG. 15.

A variety of limiting circuits are shown in Fig. 15. A dead space circuit, a back-lash and a friction circuit are shown. An understanding of their operation is left as an exercise to the reader.

RESOLVERS

An important class of function generators is that for the generation of the trigonometrical functions. Sines and cosines can be produced from specially shaped potentiometers or from linear potentiometers in which the slider is positioned by a linkage, swash plate or by other means so that the output voltage follows the sine of the angle represented by the input. Today, the most common form is the specially wound potentiometer. A continuous circular track is fed by $+V$ and $-V$ at opposite ends of a diameter, and two sliders mounted at right-angles on the centre

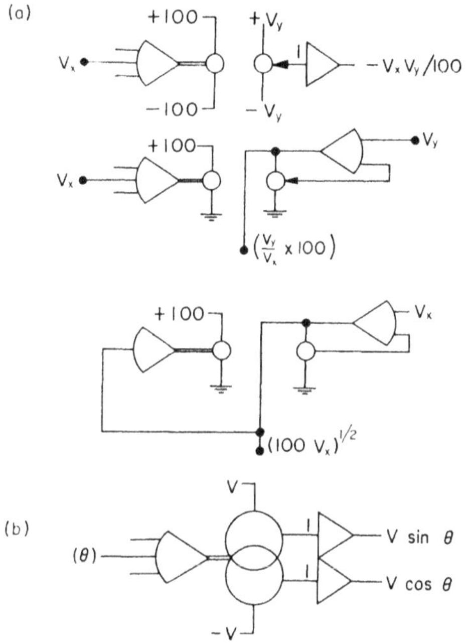

FIG. 16.

spindle give outputs of $V\sin\theta$ and $V\cos\theta$, where θ is the shaft angle. It is usual to drive the shaft by a position servo as in a servo-multiplier, with two sine–cosine and two linear potentiometer tracks. This gives a unit, which is sometimes permanently connected, for polar to cartesian or cartesian to polar coordinate transformations.

The conventional symbol is shown in Fig. 16b.

EXERCISE

Devise circuits to transform from polar to cartesian coordinates and vice versa.

NOISE GENERATORS

The normal technique for noise generation is to use a thyratron as the primary noise source. After amplification and level stabilization, a noise spectrum flat from about 30 c/s to about 3 kc/s can be produced. A region, say from 300 to 500 cs is selected by a band pass filter and shifted by a chopper relay technique, working at 400 cs, to give the sum and difference frequencies of 400 cs with the noise waveform. After filtering off the high frequencies, Gaussian noise to about 1 per cent from 0 to 35 cs, can be obtained. This is a wide enough band for most analogue uses.

OTHER FUNCTION GENERATORS

A function generator of 1 per cent accuracy can be made by placing a shaped mask in front of a cathode ray tube. The light spot is made to follow the profile of the mask by means of a photocell and an electronic position servo as described in the crossed fields multiplier, and the resulting Y-deflection voltage gives the output for any X-deflection.

A related form of function generator of about 0·1 per cent accuracy can be made from an X–Y plotter. The normal X–Y plotter is a device which positions a pen in its X- and Y-coordinates by two position servos, one driving a bar to and fro above the plotting table, the other a pen carriage along the bar. The pen can be replaced by a pick-up head which can sense the field about a conducting ink-line drawn on the paper. A radio-frequency current is then passed along the ink line. Thus, if the bar is positioned according to the X-input, the pick-up head will be driven to the line and a suitable output from the Y-servo will give the appropriate output. A two-variable function generator can be made using resistance paper with silver ink lines as equipotentials, and a pick-up probe mounted on the pen-carriage of the plotter. The output corresponding to the X and Y inputs is then interpolated between the voltages of the two nearest ink lines.

Exercises

1. Calculate the resistors for a diode function generator to give $V_0 = (V_1/10)^2$ for V_1 in the range 0–100 V. Start the first line segment from a point on the input voltage axis to give the curve near the origin. What is the accuracy of the simulation?

2. Calculate the resistors for a reciprocal function generator. Start the first line, not at the origin, but at the largest input voltage, and work back.

3. Show how to use a servo-multiplier to give squares and higher powers; show how a power series can be produced by this means.

4. The gain of an amplifier is

$$M(p) = M \cdot \frac{1}{1 + pT},$$

where $T = 20$ sec. It is used with suitable resistors to give a d.c. gain of γ. What is the gain function $\gamma(p)$? Derive $\gamma(p)$ for the more accurate value of $M(p)$ given in the text.

5. Use servo-resolvers to give the range, and azimuth and elevation angles to a target from the origin, given x-, y-, and z-coordinates. Simulate a correction for the earth's curvature.

REFERENCES

I. C. Hutcheon and D. Summers, A low-drift transistor chopper-type d.c. amplifier with high gain and large dynamic range, *Proc. I.E.E.*, March 1950.

D. L. Davies and S. F. Miles, The d.c. gain of an operational amplifier, *Electronic Engineering*, April 1962.

G. A. Goldberg, The stabilisation of wide band d.c. amplifiers for zero and gain, *R.C.A. Rev.*, 1950.

E. O. Gilbert, The design of position and velocity servos for multiplying and function generation, *I.R.E. Trans. on Electronic Computers*, Sept. 1959.

R. P. Chasmar, E. Cohen, and C. P. Holmes, The design and performance of a Hall effect multiplier, *Proc. I.E.E.*, **106,** Part B, 1959.

W. A. Scanga *et al.*, Hall-effect multipliers, *Electronics*, 15 July 1960.

E. V. Bohn, A pulse position modulation analog computer, *I.R.E. Trans. on Electronic Computers*, June 1960.

E. M. Deeley and D. MacKay, Multiplication and division by electronic analogue methods, *Nature*, 23 April 1949.

E. A. Goldberg, A high accuracy time division multiplier, *R.C.A. Rev.*, 1952.

T. R. Hoffman, Analogue multiplication using time as one variable, *Electronics*, 12 Aug. 1960.

M. V. Gal'perin, Analysis of analog computer error due to finite pass band of operational amplifiers, *Autom. & Rem. Control*, **25,** May 1964.

H. M. Paynter and J. Suez, Automatic digital setup and scaling of analog computers, *I. Soc. Am. Trans.*, **3,** Jan. 1964.

chapter 3

SCALING AND APPLICATIONS

INTRODUCTION

We have discussed in the last chapter the electronic equipment that has been produced to perform the operations of addition, multiplication, division, integration and the production of non-linear functions. But we did not discuss how this equipment was to be used in the solution of the problems for which it was designed. It can quickly be seen that the three amplifiers (two integrators and a summer) in Fig. 17 will solve the equation

$$\frac{d^2x}{dt^2} + \gamma x = 0$$

and produce a solution of the form

$$x = A\sin(\omega t + \phi)$$

where $\omega^2 = \gamma$.

Suppose that amplifier 1 has output $\ddot{x}$; then amplifier 2 gives $-\dot{x}$, and amplifier 3 gives x, so that the output of amplifier 1 is $-\gamma x$, which is equal to $\ddot{x}$, and so the equation is solved. But this *ad hoc* method must be put on an organized basis if the solution of more general equations is to be attempted. The process is best illustrated by an example.

Consider the equation describing the damped oscillatory motion

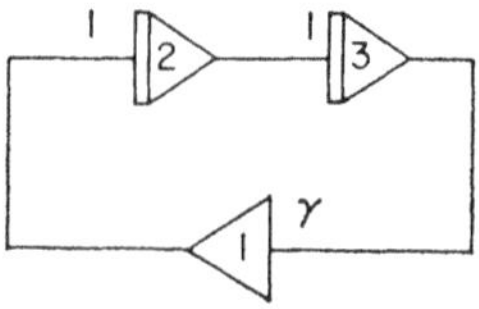

FIG. 17.

of a weight on the end of a light spring, for displacement $f(t)$ of its end

$$m\ddot{x}+\dot{x}+k[x-f(t)]=0.$$

Turn this equation round so that the highest derivative is alone on the l.h.s.:

$$-m\ddot{x}=\dot{x}+k[x-f(t)].$$

If the voltage signals representing the three terms on the r.h.s. are added into an integrator, its output will be $\dot{x}$, which can be integrated again to give $-x$, and the r.h.s. can then be produced. Thus we can quickly draw the unscaled block diagram in Fig. 18.

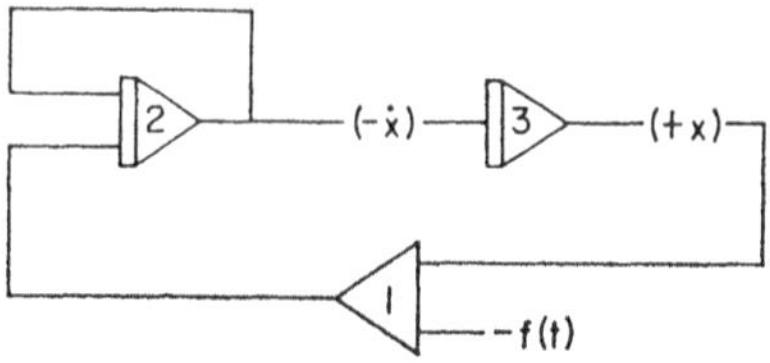

FIG. 18.

The next step is to find the values of the gains shown in the block diagram appropriate to the equation constants considered. This is the process of scaling. The physical quantity x in the equation must be related to a voltage V_x in the simulation.

SCALING

There are many different methods of scaling; they all depend on two different ways of looking at the same basic situation, i.e.

(1) $V_x = b_x . x$ volts

(2) $V_x = 100\left(\dfrac{x}{x_0}\right)$ volts

In (1), b_x has the dimensions of volts per physical unit; in (2), the physical quantity x is normalized and then multiplied by the factor 100 V. Both methods have their disadvantages in practice, the first causes muddle over the dimensions, and is cumbersome in use, the second gets mixed up on the page when writing out the equations and when drawing out the block diagrams. A method is set out here which is fairly simple to use for working out gains, and is easy to write. Of the methods used by the author, it is the most convenient, and voltages appearing in the computing machine are quickly appreciated as the physical quantity they represent. But it is recognized that a group of computer users will have their own conventions and routines. What matters is not the notation, but the efficient use of the machines available, and the speed with which an operator, skilled or unskilled, can prepare accurate equations for the machine.

SUMMER AND INTEGRATOR SCALING

Suppose the constants of the physical system of a mass and a spring give the equation

$$\ddot{x}+2{\cdot}0\dot{x}+16[x-f(t)]=0,$$

where x is the displacement in inches, and t the time in seconds. Suppose, further, that we are only interested in oscillations and changes of x where x is less than 5 in. Then, for a computer with amplifiers linear in the range ± 100 V, we can work with $(20x)$ volts to represent x in., so that when x is at its "full scale" value of 5 in., the voltage representing x, that is $(20x)$ volts, is 100 V.

We have chosen a scale for length measurements, but the equation involves rates of change also. It is therefore necessary to estimate the greatest rates of change of interest, so that the voltages representing $\ddot{x}$ and $\dot{x}$ are within ± 100 V. Now it can be seen that the equation behaves similarly to the undamped equation

$$\ddot{x}+16[x-f(t)]=0,$$

and this equation has a natural frequency of $\sqrt{16}=4$ rad/sec, so let us consider values of ratio of change equivalent to oscillation of less than 5 rad/sec; that is, consider as a first approximation that

$$x = A\sin(5t).$$

Therefore
$$\dot{x} = 5A\cos(5t)$$
$$\ddot{x} = -5 \times 5A\sin(5t).$$

We have agreed that A, the amplitude, is less than 5 in., so we represented x in. by $(20x)$ volts. From those three relations it can be seen that $\dot{x}$ is less than 25 in./sec, so that we can represent $\dot{x}$ in./sec by $(4\dot{x})$ volts. Again, $\ddot{x}$ is less than 125 in./sec^2, so that we can represent $\ddot{x}$ in./sec^2 by $(4/5\ddot{x})$ volts.

We now have voltage variables which are proportional to the original physical variables. Also they remain within the computer voltage range for the expected variations in the values of the physical variables. We require the voltage equation relating these voltage variables to be identical to the physical equation relating the physical variables.

We therefore take the original physical equation and replace its variables in turn by the voltage variables (writing them in brackets). To ensure that the equation as a whole is unaltered, every time a voltage variable is introduced in a term, any constant it introduces—within its brackets—must be removed from that term by division as a gain written without brackets.

This is illustrated, stage by stage, below.

Take each term in turn from the original equation, written out as

$$\ddot{x} + 2{\cdot}0\dot{x} + 16x = 16f(t).$$

The first term is $\ddot{x}$; our voltage variable is $(\frac{4}{5}\ddot{x})$, so the equation becomes

$$(\tfrac{4}{5}\ddot{x}) \times \tfrac{5}{4} + 2{\cdot}0\dot{x} + 16x = 16f(t).$$

Similarly, for $\dot{x}$, our voltage variable is $(4\dot{x})$, so the equation becomes

$$(0{\cdot}8\ddot{x}) \times 1{\cdot}25 + 2{\cdot}0(4\dot{x}) \times \tfrac{1}{4} + 16x = 16f(t).$$

Finally, x and $f(t)$, which are lengths, must be scaled to $(20x)$ and $20f(t)$ volts, giving

$$1{\cdot}25(0{\cdot}8\ddot{x}) + 0{\cdot}5(4\dot{x}) + 16(20x) \times \tfrac{1}{20} = 16[20f(t)] \times \tfrac{1}{20}$$

or
$$(0{\cdot}8\ddot{x}) + 0{\cdot}4(4\dot{x}) + 0{\cdot}64(20x) = 0{\cdot}64[20f(t)].$$

The process may be summarized as follows. Take the physical equation. Replace the physical variables by the voltage variables and multiply each term by an appropriate factor, such that the original equation remains. This gives the voltage equation for the voltage analogue of the physical system, and it gives the constants for that voltage equation. These constants are the gains of the amplifiers involved.

Rewriting the equation, we have

$$-(0{\cdot}8\ddot{x}) = 0{\cdot}4(4\dot{x}) + 0{\cdot}64(20x) - 0{\cdot}64[20f(t)].$$

We could, therefore, sum the terms on the right with gains of 0·4 and 0·64 to obtain $(0{\cdot}8\ddot{x})$ as the amplifier output.

If $(0{\cdot}8\ddot{x})$ is fed to an integrator with a gain of 5, the output will clearly be $-(4\dot{x})$, and, by integrating this, again with a gain of 5, the final variable, $(20x)$, is obtained. The circuit is completed by the external forcing function $[20f(t)]$, and sign changing amplifiers as appropriate.

EXERCISE

Draw out the circuit of the above, checking each sign.

It can easily be seen that the terms on the r.h.s. of the equation can be summed directly into the integrator which gives $(4\dot{x})$. The gains become, therefore, $5 \times 0{\cdot}4 = 2{\cdot}0$ and $5 \times 0{\cdot}64 = 3{\cdot}2$.

EXERCISE

Draw out this circuit. How many amplifiers are saved?

TIME SCALING

Suppose that the above circuit is set up and the gains set accurately. If the driving function is a step of 40 V, which corresponds to a step of 2 in. in the position of the end of the spring, the system will oscillate with a frequency of about 4 r/s, the oscillation dying away with a time constant of 1 sec.

This is too fast for the average recorder, though many special recorders are made for frequencies of this order. So everything must be slowed down for the purposes of the simulation. One way to do this is to replace the 1 μF capacitors by 10 μF. Another is to divide the integrator gains by 10. Both methods replace the old rates of change in the system by new rates of change one-tenth as fast, and so machine time t' and problem time t are related by the scaling equation $t' = (10t)$, i.e. 10 sec on the machine is 1 sec worth of "real" time on the physical system.

So the final gains of the time-scaled circuit are 0·2, 0·32 and 0·5 in the integrators, and 1 in the summer. These correspond to the gains of 2·0, 3·2 and 5 in the integrators and 1 in the summer in the original analogue. The final circuit is shown in Fig. 19.

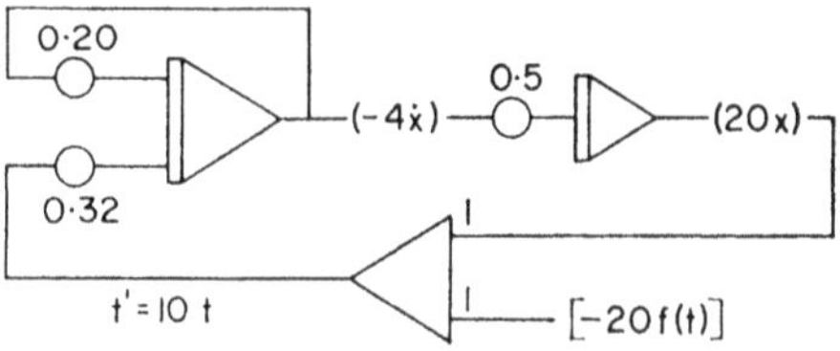

FIG. 19.

Time scaling can cause a lot of confusion. It is almost always easiest to work out everything in terms of "real" time, and then think of the capacitor rates of charge being reduced (i.e. the gains reduced or capacitors increased, in all integrators) if things change too fast for the simulation to follow. Conversely, the gains are increased if the physical system changes too slowly for convenience and accuracy on the simulation.

Record charts from a machine might, typically, have a chart speed of 3 in./min. If time scaling by a factor of 10 slower were done, as in the example, 3 in. of the chart would become a record of 6 sec of the problem solution. It is often convenient to time scale by factors of 60, so that minutes of the physical problem speed up to become seconds of the analogue, or slow down to become hours of the analogue solution. The resulting charts can then be marked appropriately in seconds or hours, though produced in minutes.

MULTIPLIER SCALING

The output from the slider of a servo-multiplier potentiometer, matched suitably, is a voltage V_z, given by

$$V_z = \frac{V_x V_y}{100},$$

where V_x and V_y are the inputs. The factor 100, which has the dimensions of volts, appears from the mechanism of the multiplier, since a multiplier multiplies by a proportion—the drive voltage V_x is normalized as the function $V_x/100$ of the potentiometer travel.

For example, suppose the coefficient of viscous damping in the term in $\dot{x}$ in the equation considered above is given by $\mu = \mu_0(1-0{\cdot}02T)$, where T is the temperature, in the range 0–100°C. Then we must generate, in place of $0{\cdot}8 \times \frac{2}{4} \times (4\dot{x})$, the term

$$0{\cdot}8\left\{\frac{2}{4}(4\dot{x}) - 0{\cdot}02 \times \frac{2}{4}\,\frac{(4\dot{x})(T)}{(100)} \times 100\right\},$$

where the factor 0·8 comes from the scaling of the rest of the equation. The scaling factors are removed as gains at each replacement of a variable by its scaled value at each stage, and the original equation of the physical system preserved, exactly as given before.

Similarly, for a division, V_x divided by V_y gives an output voltage of

$$V_z = \frac{100}{V_y} \cdot V_x$$

so the factor 100 must be removed as a gain when writing the equation for the voltages.

To summarize the scaling process, the operations involved are listed below.

(1) Decide maximum values for each variable of the physical equations, including rates of change.

(2) List physical variables and maximum values, and write down voltage variables in brackets. Each variable is multiplied by an appropriate constant to give a voltage variable which always lies in the simulator voltage range, for the values of the physical variable considered.

(3) Take each variable in turn, and substitute its voltage equivalent in the system equations term by term. Each term must be divided by the appropriate constant so that the system equations are preserved unaltered.

(4) When two physical variables are multiplied, divide by the simulator voltage range when introducing the second voltage variable. This gives the multiplier output with dimensions of volts, not (volts)^2. Now multiply by the numerical value of the voltage range so that the term is unaltered in the equation.

(5) Conversely, for a division, multiply by the voltage range, and remove this factor as a gain.

(6) If time scaling is required, after the gains have been found, reduce integrator gains by a common factor for a simulation to behave slower than the physical system, and conversely.

A WORKED EXAMPLE

The uranium fuel elements in a nuclear reactor are encased in alloy cans which are pressed on to the elements. It is found that humps can appear on the cans, and that the angular size ϕ and the normalized height x of the humps can be expressed by the equations

$$\frac{dx}{dt} = \frac{1}{16\pi^2}\phi x^2 - 1{\cdot}875\alpha, \tag{1}$$

$$\frac{d}{dt}(\phi x) = -2\phi\alpha, \tag{2}$$

where α is a parameter describing the elastic properties of the alloy can.

We first construct the unscaled diagram of the simulation circuit of Fig. 20a. Suppose that by some means we have generated $\phi\, x$ in amplifier 1. Take ϕx and divide by x to get $-\phi$ in amplifier 2. Multiply by α, and match the multiplier potentiometer for constant load in amplifier 3. This gives ϕ, which can then be integrated in amplifier 1 to give ϕx, solving equation (2). Now we know ϕx, so multiply it by x and match in amplifier 4, and then add α and ϕx^2 into an integrator, amplifier 5, to solve equation (1) for x. Two amplifiers are required to drive the x and α multipliers. Note that in this circuit, no voltage for dx/dt or $(d/dt)(\phi x)$ is generated. So these variables need no voltage equivalents.

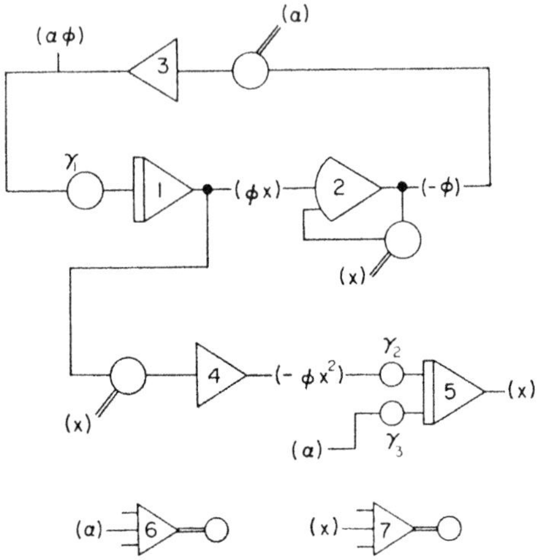

FIG. 20a.

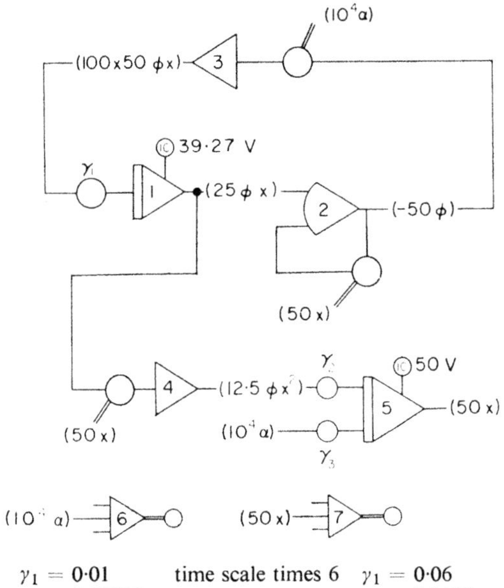

$\gamma_1 = 0{\cdot}01$ time scale times 6 $\gamma_1 = 0{\cdot}06$
$\gamma_2 = 0{\cdot}02536$ $\gamma_2 = 0{\cdot}1522$
$\gamma_3 = 0{\cdot}009375$ $\gamma_3 = 0{\cdot}05625$

Initial condition: Amplifier 1, 39·27 V or $\phi = \frac{1}{2}\pi$, $x = 1$
Amplifier 5, 50 V or $x = 1$

FIG. 20b.

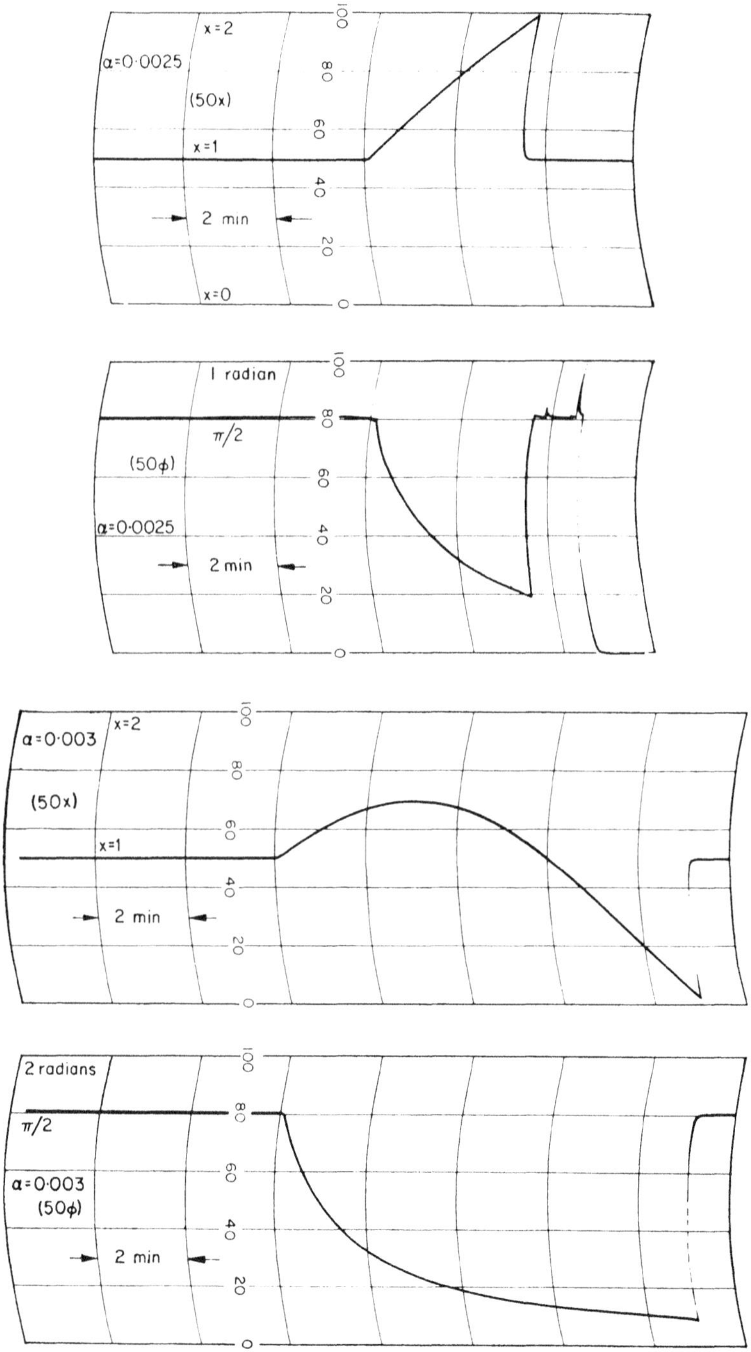
x=2
α=0·0025
(50x)
x=1
2 min
x=0
100
80
60
40
20
0
1 radian
π/2
(50ϕ)
α=0·0025
2 min
100
80
60
40
20
0
α=0·003
x=2
(50x)
x=1
2 min
100
80
60
40
20
0
2 radians
π/2
α=0·003
(50ϕ)
2 min
100
80
60
40
20
0

FIG. 20c

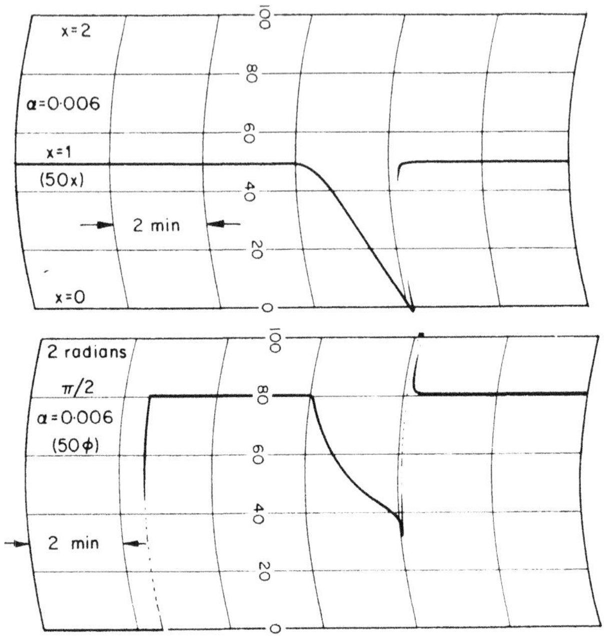

FIG. 20c

We have:

x	2 units	take $(100\frac{1}{2}x)$ volts	i.e. $(50x)$
ϕ	3 radians	take $(100\frac{1}{2}\phi)$ volts	i.e. (50ϕ)
ϕx	4 units	take $(100\frac{1}{4}\phi x)$ volts	i.e. $(25\phi x)$
α	0·01 units	take $(10^4\alpha)$ volts	

Then we have, scaling the equations,

$$\frac{1}{25}\cdot\frac{d}{dt}(25\phi x) = -2\cdot\frac{(10^4\alpha)(50\phi)}{(100)}\cdot\frac{100}{50\times 10^4}$$

$$\frac{1}{50}\cdot\frac{d}{dt}(50x) = 0{\cdot}00634\cdot\frac{100}{25\,.\,50}\cdot\frac{(25\phi x)(50x)}{(100)}-\frac{1{\cdot}875}{10^4}\cdot(10^4\alpha)$$

So $$\frac{d}{dt}(25\phi x) = -\frac{(10^4\alpha)(50\phi)}{(100)} \qquad (3)$$

$$\frac{d}{dt}(50x) = 0{\cdot}02536\frac{(25\,dx)(50x)}{(100)}-0{\cdot}009375(10^4\alpha) \qquad (4)$$

Now equation (3) is solved by amplifiers 1 and 3, whose output is:

$$-\frac{(-50\phi).(10^4\alpha)}{(100)} = (100 \times 50\phi\alpha) \text{ volts.}$$

Therefore the gain of amplifier 1 is $\gamma_1 = 0{\cdot}01$.

Similarly, the output of amplifier 4 is:

$$-\frac{(25\phi x).(50x)}{(100)} = (12{\cdot}5\phi x^2) \text{ volts.}$$

The gains of amplifier 5 which solves equation (4) are therefore $\gamma_2 = 0{\cdot}02536$ and $\gamma_3 = 0{\cdot}009375$.

The final scaled circuit is shown in Fig. 20b, and, after time scaling by a factor of 6, gave the results shown in the charts of Fig. 20c.

The analogue solution of these equations took a little over a working day. Its accuracy was limited by the computer available at the time and the presence of a dividing circuit. The results show that the hump in the can will grow without limit until the can bursts, or get narrower and flatten itself out according to the value of α.

A SMALL PROBLEM

A typical problem involving a non-linear differential equation is the determination of the closing time of the main stop and throttle valves in the steam pipes of a steam turbine. This valve is held open against a strong spring under normal conditions by oil pressure in a cylinder. If the turbine is tripped for safety reasons, the oil is rapidly drained from the cylinder, and the spring forces the valve shut against the steam pressure. The general arrangement is shown in Fig. 21.

Now if x is the travel of the piston and its area is A and a is the area of the drain port which is opened to trip the stop valve shut, the oil flow Q is given by

$$\begin{aligned} Q &= A\dot{x} \\ &= ka\sqrt{P}, \end{aligned}$$

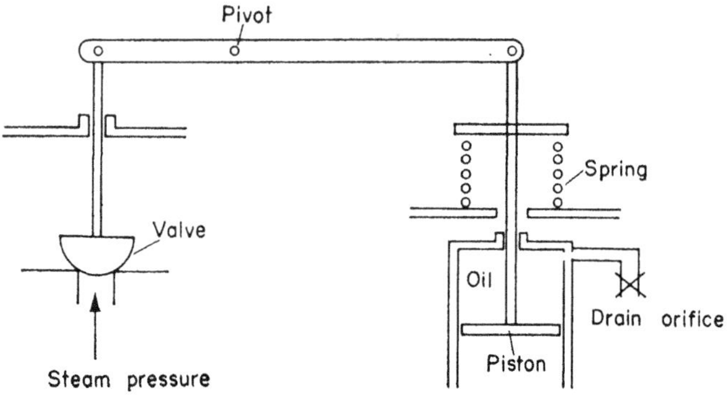

FIG. 21.

where k is a constant and P is the oil pressure in the cylinder The force on the piston is given by

$$F = AP$$

$$= A \cdot \frac{A^2 \dot{x}^2}{k^2 a^2} = 39{\cdot}16 \frac{\dot{x}^2}{a^2} \text{ lb wt.}$$

As the stop valve closes, therefore, the accelerating force is given by

$$\begin{aligned} -m\ddot{x} = &\text{ deadweight of piston} \\ &- \text{spring force} \\ &+ \text{oil force} \\ &+ \text{friction force} \\ &+ \text{steam pressure force on the valve.} \end{aligned}$$

A typical set of constants is given below:

Oil-piston diameter	11 in. outside, 3·75 in. inside
Oil-piston travel	1·625 in.
System mass	821 lb
Deadweight force	767 lb assisting opening
Friction	440 lb resisting closure
Steam force	1950 lb assisting opening
Spring force: closed	3422 lb assisting closure
open	5614 lb assisting closure
Spring rate	1350 lb/in.

It is left as an exercise to show that the equation of motion of the system is

$$2\cdot115\ddot{x} = -39\cdot16\frac{\dot{x}^2}{a^2}+1350x+265.$$

Taking maximum values,

$x_m = 2$ in. work in $(50x)$ volts,
$\dot{x}_m = 50$ in./sec work in $(2\dot{x})$ volts,
$\dot{x}_m^2 = 2500$ (in./sec)^2 work in $(0\cdot04\dot{x}^2)$ volts,

this gives the scaled equation

$$d(2\dot{x}) = -\frac{39\cdot16}{a^2}\cdot\frac{2}{2\cdot115}\cdot\frac{1}{0\cdot04}(0\cdot04\dot{x}^2)$$
$$+\frac{1350}{2\cdot115}\cdot\frac{2}{50}\cdot(50x)+\frac{265}{2\cdot115}\times\frac{2}{100}(100)$$
$$= -\frac{925\cdot9}{a^2}(0\cdot04\dot{x}^2)+25\cdot54(50x)+2\cdot506(100).$$

If the simulation is now slowed down by a factor of 500, the gains involved and the rate of stop valve movement come into the range of a feasible simulation. The circuit is given in Fig. 22.

It is then a simple matter to take runs from the simulator for various values of drain valve area. Figure 23 shows a typical run

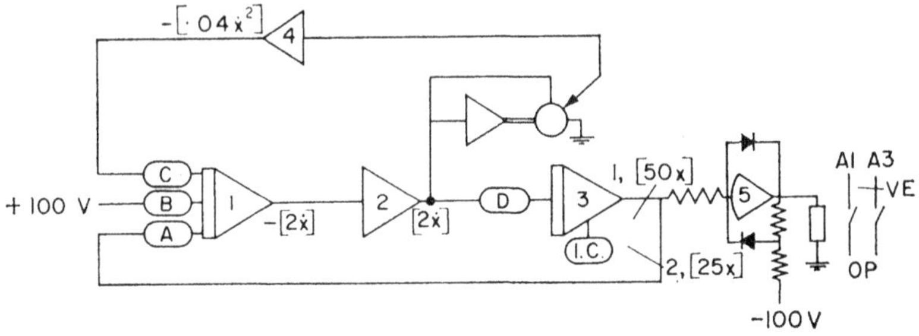

Coeff.	Throttle
A	0·0512
B	0·005
C	According to orifice size
D	0·05

FIG. 22.

giving valve travel and velocity for a 1·75in. diameter drain orifice. From such runs, the orifice giving optimum closing time without damage to the steam-valve seat due to excessive closing velocity on impact can be chosen. A laborious hand calculation

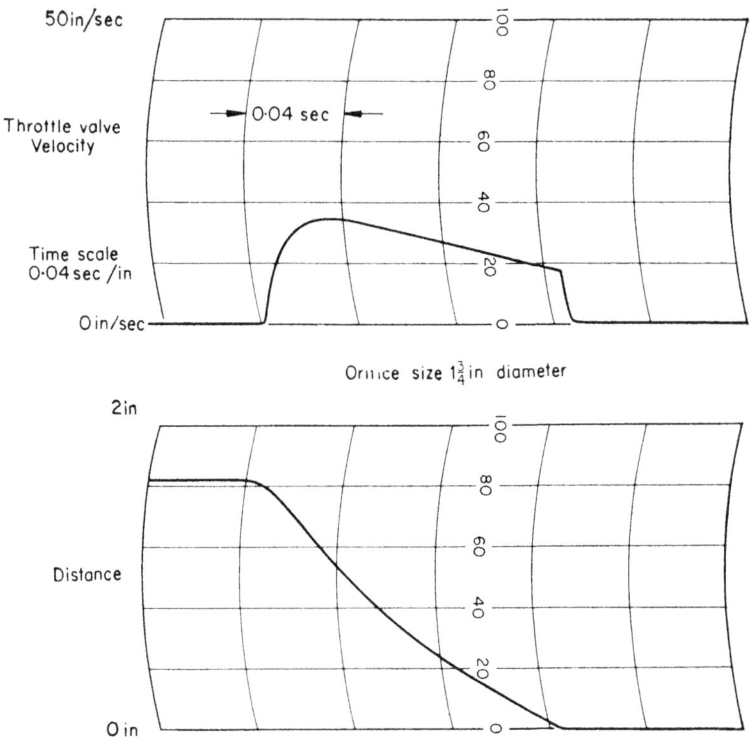

FIG. 23.

is otherwise needed to work out step by step the movement of the valve. This type of calculation becomes a simple routine job for an analogue computer, with the possibility of useful answers for the designer by return of post if computer time is available.

ESTIMATION OF MAXIMUM VALUES

A problem is usually scaled wrongly once or twice before its completion. The maximum values must be estimated from considerations of the physical system and from the solutions or estimated solutions of the equations. With many systems of importance this is comparatively easy, since some steady state will be known. But for non-linear systems the cut and try method is often the only way, and far quicker than involved analysis of the equations.

METHODS OF SOLUTION

It is in general an easy matter to draw out a circuit diagram for the simulation of a set of differential equations on the lines indicated in the two examples given. However, a more explicit classification of the methods used is in order.

The "direct method" of solution is used to solve for functions of the independent variable, simulated as time, from the equations. For example, the coordinates of a projectile fired with velocity V at an angle of elevation θ are

$$x = V\cos\theta\,.\,t,$$

$$y = V\sin\theta\,.\,t - \tfrac{1}{2}g\,.\,t^2.$$

So x and y can be generated direct from integrators producing $V\cos\theta\,.\,t$ and $V\sin\theta\,.\,t$ from a constant voltage, and an integrator to give a ramp $\frac{1}{2}g\,.\,t^2$ for the integrator for y.

The "indirect method" of solution is that used in the two examples given earlier. It was first proposed by Lord Kelvin when solving a second-order differential equation with his brother's ball-and-disc integrator. He fed the input of his machine with a trial guess to the solution and recorded the output. This output was fed to the input as a second guess, and the solution produced by an iterative process on his open-loop machine. He then conceived the notion of connecting the output to the input, that is, of closing the loop, and to his delight the solution was obtained immediately. This step is quite obvious to those who have experience of feedback systems, but can be puzzling when met for the first time. The indirect method, Kelvin's method, is the normal line of attack on a circuit.

Often a problem variable can be generated by the solution of a subsidiary or implicit equation. The dividing circuit of Fig. 24g solves an implicit equation

$$V_1 + \frac{V_0 V_2}{100} = 0.$$

This method is often simpler and more accurate than generating a variable with a diode function generator. An example is the generation of the exponential $y = e^{-x}$. Since $dy/dt = -y(dx/dt)$, the reader may quickly draw a circuit to give y knowing dx/dt, and the initial conditions.

An extension of the implicit function method is the use of amplifiers without feedback. If, for example, (x, y) coordinates are to be converted to polars (r, θ) then $r = x\cos\theta + y\sin\theta$, and $r = (x^2 + y^2)^{\frac{1}{2}}$. If these terms are generated with a servo-resolver and with multipliers, and added with opposite signs into an amplifier without feedback, the output of this amplifier may be taken as θ and used to drive the servo-resolver. When the two values of r agree, the resolver stops turning, and a solution for θ results.

This technique has been used, for example, to simulate the behaviour of a boiler which first heated up the water as it passed through the system, then it boiled at a constant temperature dependent on the rest of the system. No explicit equation for the point where boiling started could be found, so the multipliers which fixed that point were driven from an amplifier with very little feedback (a gain of about 500) by comparing two values of the steam flow from the system. These values were equal if the equations were satisfied, and a solution was formed on the system by reducing the difference to a very small value.

When a problem is scaled, it may happen that the gains of some of the integrators come out to be far too large for simulation. For example, the time constant for conduction through a tube between water and hot gas in a heat exchanger might be many times that of the gas itself. Consider the equation for the heat balance of the gas in a section of such a heat exchanger of length 1 unit. Then

$$m_g c_g \frac{dG}{dt} = Mc_g(G_i - G_0) - H(G - T),$$

where $m_g c_g$ is the thermal capacity of the gas, M is the gas mass flow, H is a constant, G is the mean gas temperature between inlet G_i and outlet G_o, and T is the tube temperature. If the gas time constant (its thermal capacity) is very small, that is the gains of an integrator giving G would be very large, we may write, approximately,

$$Mc_g(G_i - G_o) = H(G - T).$$

So the gas temperature can be generated with a summer. As another example, when considering the changes in power from a nuclear reactor due to the daily variation of load, the time constants of the fuel rod and can, which are about 2 sec, can be neglected.

SOME STANDARD CIRCUITS

Figure 24 shows some standard circuits which may be regarded as the blocks from which to form a simulation circuit. The summer, integrator and lag are straightforward. Figure 24d is a lag-lead circuit which can be used as an approximate differentiator with $\gamma_2\gamma_3 = \gamma_1\gamma_4$. The use of positive feedback to achieve a perfect differentiator when $\gamma = 1$ is shown in Fig. 24f. The dividing circuit shown gives a square root if the multiplier drive is the output.

It is sometimes necessary to simulate a transport delay. These give an equal time lag to signals of all frequency, with no loss of amplitude, and have transfer function given by

$$\bar{V}_0 = e^{-pT}\bar{V}_i, \text{ where } T \text{ is the delay time.}$$

This cannot be simulated exactly with summers and integrators. Some units have been designed which produce a delay with magnetic tape by setting the read and write heads at a distance apart, or with curve following devices on the same principle. But the most common approximations are derived from the Padé expansion of e^{-pT} and are:

$$e^{-pT} \approx \frac{1 - 0{\cdot}5Tp}{1 + 0{\cdot}5Tp}$$ delay = T sec, maximum phase shift 45° (5 per cent error).

and $$e^{-pT} \approx \frac{12 - 6pT + p^2T^2}{12 + 6pT + p^2T^2}$$ delay T sec, maximum phase shift 155° (5 per cent error).

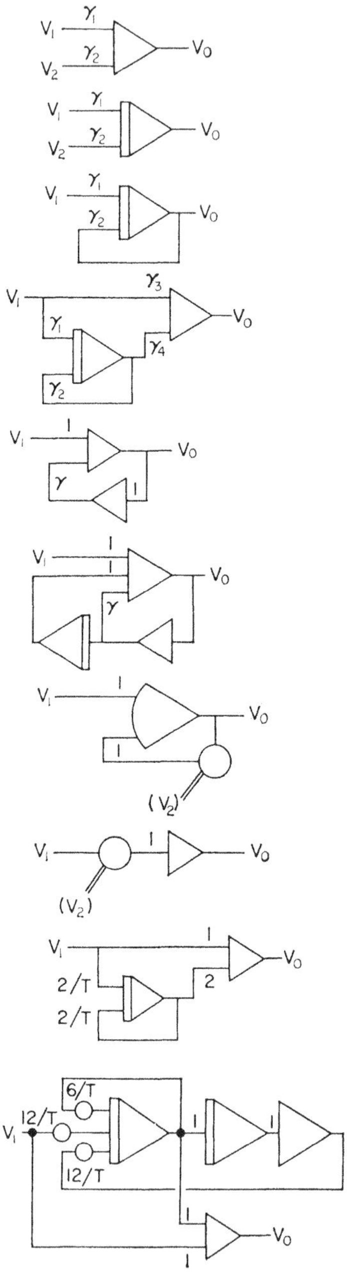

(a) $$-V_0 = \gamma_1 V_1 + \gamma_2 V_2$$

(b) $$-V_0 = \gamma_1 \frac{V_1}{p} + \gamma_2 \frac{V_2}{p}$$

(c) $$-V_0 = \frac{\gamma_1}{\gamma_2 + p} V_1$$

(d) $$-V_0 = \gamma_3 V_1 - \frac{\gamma_1 \gamma_4}{\gamma_2 + p} V_1 = \frac{\gamma_2 \gamma_3 - \gamma_1 \gamma_4 + \gamma_3 p}{\gamma_2 + p} V_1$$

(e) $$-V_0 = \frac{V_1}{1 - \gamma}$$

(f) $$-V_0 = \frac{p V_1}{1 + (1 - \gamma) p}$$

(g) $$-V_0 = 100 \frac{V_1}{V_2}$$
if $V_2 = V_0$,
$$-V_0 = 100 \sqrt{V_1}$$

(h) $$-V_0 = \frac{V_1 V_2}{100}$$

(j) $$-V_0 = \frac{1 - \frac{1}{2} T p}{1 + \frac{1}{2} T p} V_1$$

(k) $$-V_0 = \frac{12 - 6pT + pT^2}{12 + 6pT + pT^2} V_1$$

FIG. 24. (j) First-order finite delay of time T. (k) Second-order finite delay of time T.

This transfer function corresponds to the equation

$$\frac{x(t)}{y(t)} = 1 - \frac{12pT}{12+6pT-pT^2};$$

but
$$-\frac{12pT}{12+6pT-pT^2}\cdot y = v;$$

therefore
$$x = y+v;$$

therefore
$$\ddot{v}+\frac{6\dot{v}}{T}+\frac{12v}{T^2} = -\frac{12}{T}\cdot y.$$

The circuits for these approximations are given in Fig. 24.

EXERCISES

1. Give the gains and circuit of an oscillator to produce an a.c. signal of variable amplitude with frequencies from 1 c/s down to 1 cycle in 10 min, using standard integrators and summers. How does the leakage resistance of the integrator capacitors affect the performance of the oscillator? How could this resistance be estimated? If possible, devise a circuit to compensate for this loss continuously, and test the circuits on a computer.

2. Give a circuit which will give the in phase and quadrature terms of the output of a system driven by a perfect oscillator by generating the first two terms of the Fourier series expansion of the output.

3. Give circuits to transform polar coordinates to cartesian coordinates and vice versa, using sine–cos resolvers or function generators or other means.

4. Investigate the equation $\ddot{x} + 2\zeta\omega_n\dot{x} + \omega_n{}^2x = f(t)$ on an analogue computer with sinusoidal and step inputs. Show the phenomena of resonance, critical damping, etc.

5. Write down the equation of motion of a simple pendulum of length a. Draw scaled circuits assuming $\sin\theta = \theta$, $\sin\theta = \theta - \frac{1}{6}\theta^3$, and then solve the exact equation by transforming to (x, y) coordinates and using $\dot{x} = a\cos\theta\,.\,\dot{\theta}$, $\dot{y} = -a\sin\theta\,.\,\dot{\theta}$. How does the period vary with amplitude?

6. Show that the simultaneous linear equations

$$\sum_j a_{ij}x_j - c_i = 0$$

where i, j take the values 1, 2, . . . , n, may be solved by (1) generating x_i from the ith equation in a summer, (2) solving the equations $\sum_j a_{ij}x_j - c_i = \mathrm{d}x_i/\mathrm{d}t$ with integrators.

Bode Plot	Transfer Function	Constants	Gain
		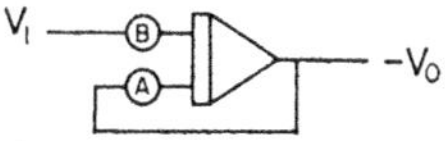	
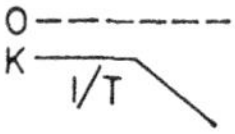	$\frac{K}{1 + pT}$	$T = \frac{1}{A}$ $K = \frac{B}{A}$	$A = \frac{1}{T}$ $B = AK$
		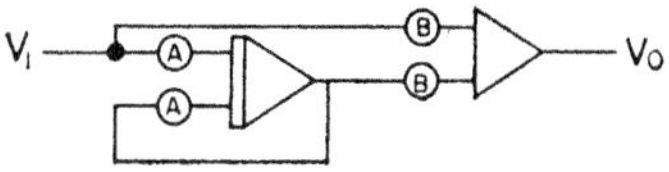	
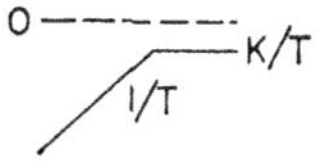	$\frac{Kp}{1 + pT}$	$T = \frac{1}{A}$ $K = BT$	$A = \frac{1}{T}$ $B = AK$
		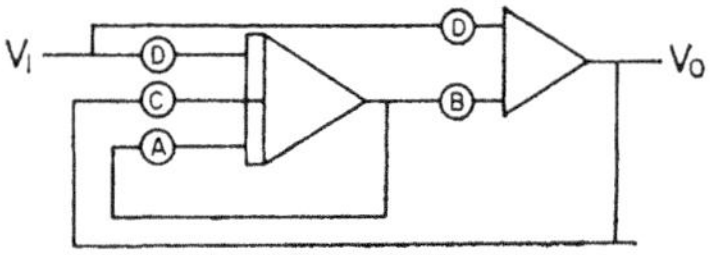	
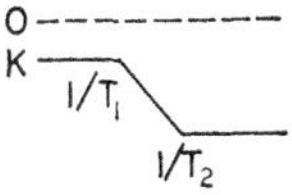	$\frac{K(1 + pT_2)}{(1 + pT_1)}$	$T_1 = \frac{1}{A - BC}$ $T_2 = \frac{1}{A - B}$ $K = \frac{T_1}{T_2} D$	$A = B + \frac{1}{T_2}$ $C = \frac{T_1 - T_2}{BT_1T_2} + 1$ $D = K \frac{T_2}{T_1}$
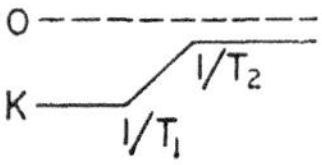	$\frac{K(1 + pT_1)}{(1 + pT_2)}$	$T_1 = \frac{1}{A - B}$ $T_2 = \frac{1}{A}$ $K = \frac{T_2}{T_1} D$	$A = \frac{1}{T_2}$ $B = \frac{T_2 - T_1}{T_1T_2}$ $C = 0$ $D = \frac{KT_1}{T_2}$

FIG. 25.

It may be shown that a solution is not necessarily obtained if the matrix of coefficients (a_{ij}) is not positive definite. A solution is always obtained from method (2) with

$$\sum_j a_{ij}x_j - c_i = \frac{dy_i}{dt},$$

$$\sum_j a_{ji}y_j = \frac{dx_i}{dt}.$$

7. By sketching circuits, find the roots of the equation

$$P_n(x) = x^n + a_{n-1}x^{n-1} + \ldots + a_1x + a_0 = 0$$

by writing

$$P_n(x) = \{x[x(x \ldots (x + a_{n-1}) \ldots + a_3) + a_2] + a_1\} + a_0.$$

Generate $P_n(x)$ with multipliers and find the roots by varying the input. Hence scale and solve

$$P_4(x) = x^4 + x^3 - 19x^2 + 11x + 30 = 0$$

(roots -5, -1, 2, 3).

It can be shown that, to obtain useful results when solving a polynomial, much higher precision is required in the working than is given in the resulting root (see *Modern Computing Methods*, National Physical Library, pp. 59–60). It is therefore normally impracticable to solve matrices or polynomials above the third or fourth order on analogue computers.

8. Draw circuit diagrams for the equations:

$$\ddot{x} - \varepsilon(1 - \tfrac{1}{3}\dot{x}^2)\dot{x} + x = 0 \quad \text{(Rayleigh's equation),}$$
$$\ddot{x} + k\dot{x} + (A + Bx^2)x = F(t) \quad \text{(Dufferin's equation),}$$
$$\ddot{x} - \varepsilon(1 - x^2)\dot{x} + x = 0 \quad \text{(Van der Pol's equation),}$$
$$t^2\ddot{x} + t\dot{x} + (t^2 - n^2)x = 0 \quad \text{(Bessel's equation).}$$

9. Derive the circuits and transfer functions shown in Fig. 25.

10. A servo-controller has transfer function

$$\frac{\theta_0}{T_e} = K_1 \frac{(1 + 30p)}{p(1 + p)} \cdot \frac{(1 + 13p)}{(1 + 1{\cdot}3p)}.$$

Draw a simulation circuit, and scale for

$$|T_e| < 25°\text{C}, \quad 0 < \theta_0 < 1.$$

Add a limit so that the rate of change of output has a limit given by $|p\theta_0| < 1{\cdot}6$.

chapter 4

COMPUTERS AND ASSOCIATED EQUIPMENT

GENERAL DESIGN OF A COMPUTER

A computer user, when studying a machine for the first time, will discuss the specification of the amplifiers and multipliers, and the provision made for other equipment such as function generators, relay amplifiers, etc. A user wants a simple, clear patchboard, accurate means of gain setting, and simple control over the computer. Amplifier gains should be standard, or standard except for a few selected amplifiers which can have special components wired in. An adequate number of amplifiers must be available for the type of problem the user is solving; and problems have a habit of growing as additional complications are added. Maintenance and servicing of the machine should be simple and rapid. And, of course, the price must be low.

PATCHBOARD

A typical modern computer of more than about thirty amplifiers will have a central patching bay, with a removable patchboard for making the amplifier interconnections. Problems can be set up away from the machine on the patchboard, and, once tested, can be stored for future use. The patchboard usually groups a set of amplifiers together, with from five to ten inputs and four or five output sockets for each amplifier. These groups will be separated by potentiometers, sets of commoned sockets joined together, reference voltage sockets, and multiplier potentiometers or other multiplier sockets. Other sockets will give lines to recorders, switches, relay contacts, diodes, or to outlet sockets which may be joined to another computer working in conjunction with the first. The sockets are usually marked and colour coded to simplify patching.

CONTROL

The control panel of a computer will give operating conditions for "compute", "hold", "stand-by", "gain set" and so forth. Control is via push-buttons or switches, operating relay interlock systems which in turn work relays on the amplifiers to give the appropriate computer conditions. Lights indicate the condition the computer is in at any time. Some computers can be operated repetitively, that is, the problem set on the machine is solved once every second, or every minute, or every time a certain variable reaches a given level. Another repetitive mode of solution resets the machine at, say, fifty times a second, and the solution is displayed on an oscilloscope. Usually two or more machines can be coupled together and controlled from one central desk. In this way, a basic computer of say, fifty amplifiers, can be used with two others to give a computer of 150 amplifiers.

MEASUREMENTS

A high impedance digital voltmeter giving a reading of voltage to four figures with sign is usually incorporated in a computer. Alternatively a sensitive voltmeter can be fitted. By balancing for zero voltage difference between the slider of a calibrated potentiometer or a decade resistance box, and the unknown voltage, accurate measurements can be made without loading that unknown voltage. For gain setting, either a known voltage can be connected to any input whilst all other inputs are earthed, and the output of the amplifier monitored as the gain is set to the required value with the potentiometer, or the slider of the potentiometer can be monitored, taking care not to load it with the voltmeter, and set to give the required fraction of the input voltage. In the first case, relays make all integrators into summers in the gain set condition of the computer. Thus the gains of integrators are set up exactly as are summers. If the potentiometer sliders can be read direct, this is not necessary, as the voltage read at the loaded slider gives the gain directly.

RECORDS AND READOUT

The normal method of recording the results of a computer run is on strip chart recorders. A moving coil meter movement drives a pen or a hot wire across a moving paper chart to give a continuous record of the variable. Alternatively, a pen is positioned over a sheet of graph paper in the X- and Y-directions by two servos driven by two machine variables, or arranged to give a record with the X-direction driven by a time base. Some computers have high speed automatic print-out of the voltages on all amplifiers. This printed record can be kept with the chart records of the runs, or used as a guide for checking the accuracy of the machine and the gain setting.

OVERLOAD INDICATIONS AND ALARMS

In order to ensure that the computer is operating within its voltage range, a buzzer or flashing light is arranged to show when any amplifier is outside its operating range. A convenient method is to trigger a neon light from the output of the a.c. channel of the amplifier, which produces large pulses when the virtual earth is lost due to excessive input voltage or too high output current, or other misuse of the amplifier. Similar alarms show when a multiplier is being driven off the end of the potentiometer track or outside its operating range. The power supplies and voltage reference supplies will normally have some form of overload protection, and some form of automatic sequencing will ensure that the computer is turned on and off in the right manner.

MAINTENANCE

The maintenance of a computer should not normally worry the operator. A routine maintenance scheme, and a full-time maintenance section are necessary for large installations, and for a smaller computer, faulty equipment is often maintained by the manufacturers.

SOME TYPICAL COMPUTERS

THE PACE SYSTEMS TR-10

This computer is a compact transistorized desk instrument for the solution of small problems involving up to twenty amplifiers. The machine unit is 10 V, which leads to an accuracy of 0·1 per cent with the high quality resistors and capacitors used in the machine. Twenty coefficient potentiometers are supplied, and multiplication is performed with diode quarter-square networks. The amplifiers are patched into the circuit required using plug-in resistors and patchleads to capacitor units and potentiometers. The arrangement of the computer can be seen in Fig. 26; the amplifiers are at the foot of the vertical panel with the capacitor units, the diode multipliers and function generators, reference voltage points, comparitor relays, switches and common points appearing above them in plug-in units. At the top of the panel are the potentiometers and their patching sockets. The control panel is extremely simply arranged. A three-position switch gives reset, hold and operate conditions of the computer. Gain setting is done with an accurate graduated potentiometer and a multi-range meter arranged in a null circuit. A multi-position switch gives the output of each amplifier and the values of the power and reference voltage supplies. Any amplifier overload is indicated by a light on the panel. The computer is turned on by a single main switch, and uses about 60 W.

THE SOLARTRON SC-30 (FIG. 27)

This machine incorporates all the features of many much larger computers. It is a thirty-amplifier unit, which can be used in conjunction with other computers to form a much larger machine. Twenty of the amplifiers are sum/integrate units with five inputs, values $3 \times 1\,\mathrm{M}\Omega$, $2 \times 0{\cdot}1\,\mathrm{M}\Omega$ (0·1 per cent), and four feedback components of $1\,\mathrm{M}\Omega$, $0{\cdot}1\,\mathrm{M}\Omega$, $1\,\mu\mathrm{F}$, and $0{\cdot}1\,\mu\mathrm{F}$ (0·1 per cent). The remaining ten amplifiers are summers with the same range of gains. Sixty coefficient potentiometers are provided. Four servo-multipliers or various combinations of mark-space and height units of an accurate electronic multiplier can be fitted.

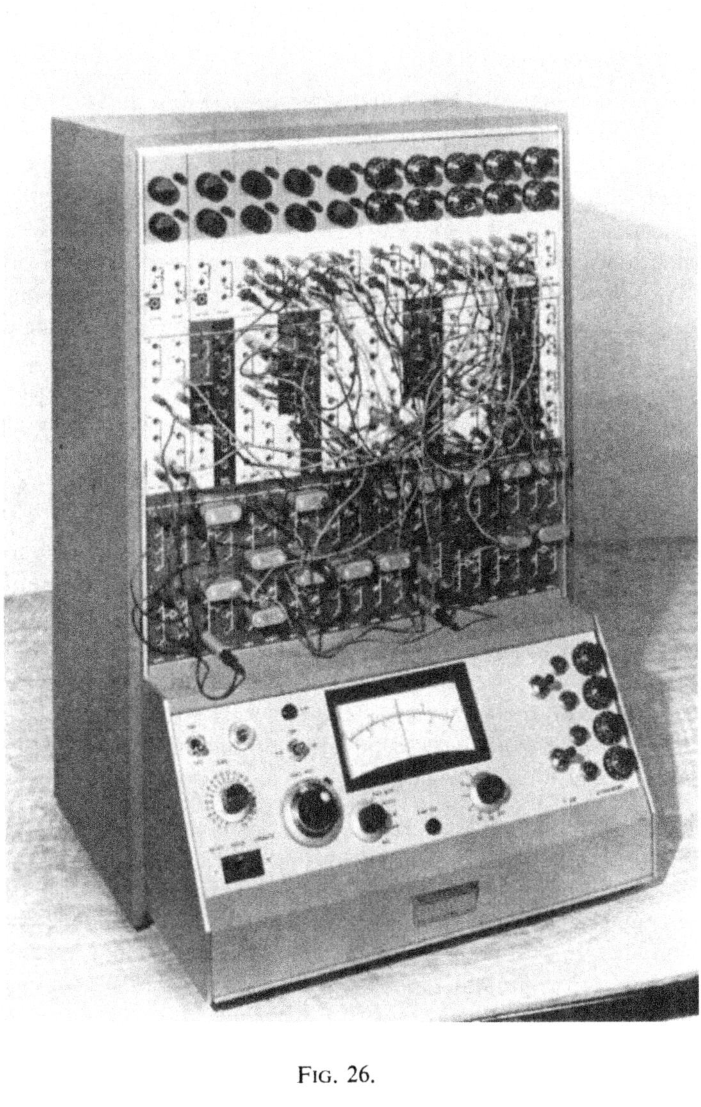

FIG. 26.

The computer has a removable patchboard with amplifiers, multipliers and standard limiters, function generators, relay comparitors, and 100 V reference supplies simply laid out and colour coded on the patchboard. A five-digit digital voltmeter is provided for gain setting and voltage measurement, but a meter and nulling potentiometer can be used if preferred. Push-buttons set the

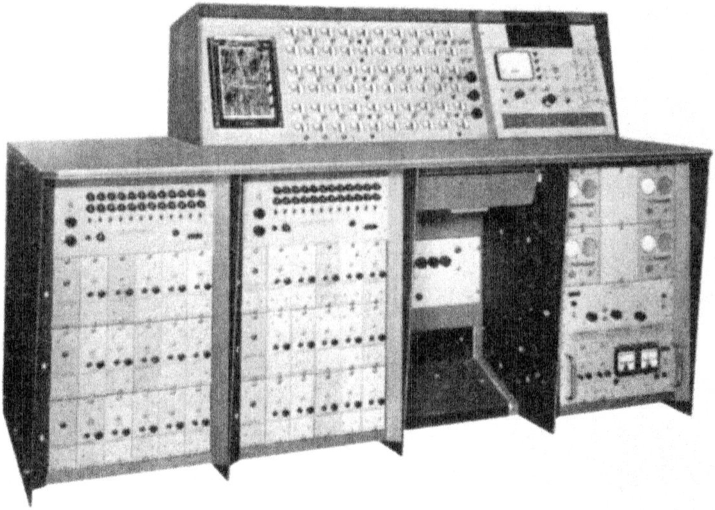

FIG. 27.

computer to potentiometer set, problem check, compute and hold conditions, with indicating lights showing the computer state. The computer can be arranged to work in a repetitive mode, either resetting itself or resetting at fixed intervals. It can also be arranged to work with one part of the computer controlling the other, via the reset relay lines which appear on the patchboard, or to give a voltage scale change by a decade. Five amplifiers can be used in a non-standard manner, with any circuit wired round them on the special component trays provided. The whole is housed in a desk unit 80 in. × 36 in. high × 38 in. deep, with power supplies and adequate desk surface.

THE PACE SYSTEMS 231-R (FIG. 28)

Electronic associates have pioneered high accuracy analogue computers, and, from their standard range, a computer can be built up to fit the prospective users' requirements. A typical Pace computer might have 100 amplifiers, 150 potentiometers, 20 or 30 multipliers or resolver units, 20 diode function generators and the associated racks, power packs, digital voltmeter and display equipment. The computer has an all-metal removable patch-board, the interconnections being made by shielded patching wires which considerably reduce the circuit noise. The unit might have automatic voltage print-out, $X-Y$ plotters, a maintenance bench and equipment, and perhaps two-variable function generators, electronic multipliers, and other equipment from the manufacturers range.

The unit can be extended by the use of the interconnecting trunk wiring to give a much larger unit, controlled from the central desk. A wide range of amplifiers, multipliers, servo-multipliers and resolvers, electronic multipliers, function generators, $X-Y$ plotters and recorders is available. A notable feature is the automatic potentiometer setting system, ADIOS (Automatic Digital Input-Output System). A paper tape can be prepared, which will set all the gains of a problem in a matter of minutes by means of servo-motors on the potentiometers. With its aid, a problem can be taken off a machine and another put on, ready for production, in 20 min or so. This overcomes the major inflexibility of most analogue machines, since the inevitable machine idle time which occurs during a problem solution can be used for other problems, and the time booking system usual on digital machines can be used.

The standard checking, gain set, computing and problem hold facilities are available, using a push-button control system, for real time solutions. Alternatively, the machine can be used repetitively at from 10 to 50 c/s of problem solution, using a separate display. The cabinet containing all computing elements is thermostatically controlled to reduce any possible drift resulting from temperature changes. The units slide out of this cabinet for maintenance, and replacement units can be plugged in at once.

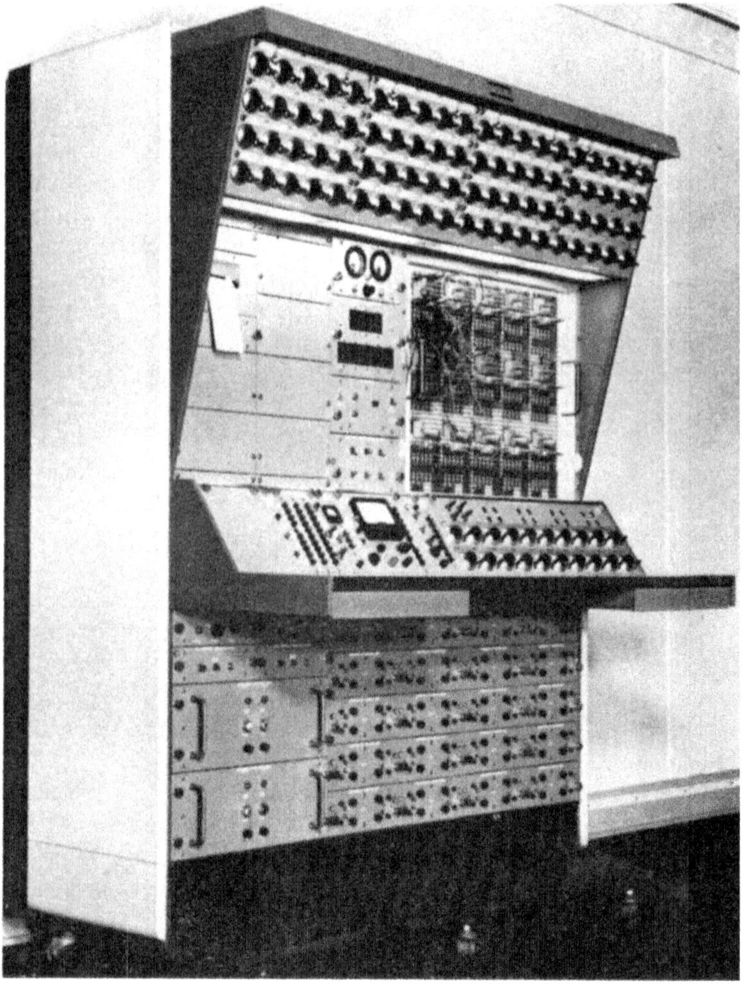

FIG. 28.

The high standard of the Pace system is indicated by the manufacturers' claim that they have supplied 70 per cent of the world's analogue installations.

THE ENGLISH ELECTRIC SATURN (FIG. 29)

Saturn was designed for the simulation of nuclear power station plant with the simulation of a complete nuclear station in mind. It has six computers of 252 amplifiers which can be used together to form a single unit of 1512 amplifiers, and a large number of multipliers and non-linear devices. It is permanently housed in the end of a laboratory building, with the amplifier and component racks together with the power supplies and the maintenance facilities upstairs, and the computer control room with the six individual control panels, recorders and patchboard units on the ground floor below. A single computer comprises 252 amplifiers divided into twelve racks of twenty-one amplifiers, and 600 potentiometers, fifty per rack. High-performance servo-multipliers are used, there being three available per rack.

A detachable patchboard, about 2ft square, is subdivided into twelve sections, one section of which is shown in Fig. 30.

The functions of the various amplifiers, and their gains, are shown on page 83. Thirteen of the twenty-one amplifiers can be used as integrators or summers, three are for use with the multipliers, two are for use as function generators, comparison relays, or for other special uses, and the remaining three are summers.

The multipliers and potentiometers are indicated on the diagram. The control panel has a subsidiary patchboard connected to the main patchboard for use with the switches and voltage-source potentiometers on the panel. The outputs of every amplifier are brought to a telephone-type jacking board between the amplifier and the patchboard socket of its output; this enables voltages to be injected using jacks at any point of the circuit when setting up or during runs on the machine. Control push-buttons on the desk give clamp, operate, gain set, operate in one-tenth time, set initial conditions, zero set, and other computer conditions. The gain-setting potentiometers are in a short corridor adjacent to the control desk, and gains are set using the computer digital

FIG. 29.

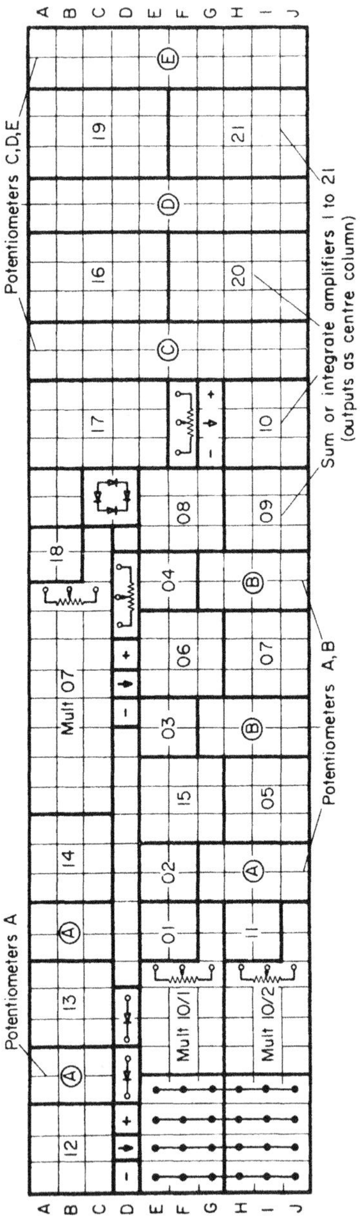

FIG. 30. Saturn patchboard layout for one rack of equipment.

Amplifier
number

01 Multiplier drive amplifier.
02 Summer gains of 1, 10, 10.
03 Summer gains of 1, 10, 10.
04 Summer gains of 1, 10, 10.
05 Summer/integrator gains of 1, 1, 1, 10, 2 (for multiplier) Initial Condition.
06 Summer/integrator gains of 1, 1, 1, 10, 2 (for multiplier) Initial Condition.
07 Summer/integrator gains of 1, 1, 1, 10, 2 (for multiplier) Initial Condition.
08 Summer/integrator gains of 1, 1, 1, 10, 2 (for multiplier) Initial Condition.
09 Summer/integrator gains of 1, 1, 1, 10, 2 (for multiplier) Initial Condition.
10 Function generator or special purpose amplifier.
11 Multiplier drive amplifier.
12 Summer/integrator gains of 1, 1, 1, 10, 2 (for multiplier) Initial Condition.
13 Summer/integrator gains of 1, 1, 1, 10, 2 (for multiplier) Initial Condition.
14 Summer/integrator gains of 1, 1, 1, 10, 2 (for multiplier) Initial Condition.
15 Summer/integrator gains of 5×1; 2×10; 2×2 (for multiplier) Initial Condition; or one-tenth of these values at Integrator.
16 Summer/integrator gains of 5×1; 2×10; 2×2 (for multiplier) Initial Condition.
17 Function generator or special purpose amplifier.
18 Multiplier drive amplifier.
19 Summer/integrator gains of 1, 1, 1, 10, 2 (for multiplier) Initial Condition.
20 Summer/integrator gains of 1, 1, 1, 10, 2 (for multiplier) Initial Condition.
21 Summer/integrator gains of 1, 1, 1, 10, 2 (for multiplier) Initial Condition.

[− | ⏚ | +] : −100 V, signal earth, +100 V sockets.

A0 to A9, B0 to B9 . . . E0 to E9 are the references of potentiometers.
Row D gives lines to the control panel auxiliary patchboard.
Diodes for limiters are shown in rows C and D.

voltmeter. Covers prevent subsequent movement of the potentiometer dials by mischance. A panel shows when any amplifier has overloaded, and another panel below indicates whether the multipliers are operating correctly. Twelve recorders are mounted on the panels above the desk, and other recorders can be used if desired.

The whole computer is serviced by maintenance staff, who also turn on and check the computer every morning.

Operating experience on this machine has shown that machine outage can be as low as 5 per cent. Typical usage of the machine is, however, 40 per cent production time, 40 per cent problem changing time, 10 per cent idle time, 3 per cent wasted due to incorrect usage, and 7 per cent fault time.

THE ROYAL AIRCRAFT ESTABLISHMENT TRIDAC

Tridac was designed for the simulation of aircraft and missile systems, although it can be used for any other type of analogue problem. It was among the first large computer systems designed in this country, and is one of the largest in the world. However, its design and performance seem in many respects out of date by modern standards, and it is chiefly of historical interest here.

The computer has about 650 amplifiers, operating to a machine unit of 30 V, and a number of servo-multipliers and resolvers. It is housed in a special two-storey building at R.A.E., Farnborough. The amplifiers and other electronic units are mounted in cabinets which divide up into eleven "rafts" which vary in detail. Each raft has a patching unit, which is connected to the central control and patching desk. The multipliers and resolvers which are hydraulically operated are in an annexe with their power supplies.

The whole computer can be controlled in subsections, and this control can be patched on the control desk. Various repetitive computing modes are available, and standard X–Y recorders, oscilloscopes and strip recorders give the output of the computer. Overload alarms and indications are given at the desk and on the rafts. A regular maintenance system ensures maximum reliability.

The basic design of the computer started in about 1950, and it was being operated by about 1954. Since then it has been engaged on a wide variety of missile and aircraft problems. Modifications to the original design of the computer have made it more versatile; it was originally built as permanently connected units for the simulation of the aerodynamics, the axis transformation, radio system and so forth. This type of layout was widely used in early computers, and has been almost entirely replaced by the more flexible arrangement of central patching with every amplifier available for any use.

There are many other commercial computers, and there is not space to describe more than the few of them mentioned. The ones mentioned are fairly well known, typical, and up to date, but are not specifically recommended.

EXERCISE

As an exercise for the student, or as a project for a group, the design of a computer of, say fifty amplifiers is recommended. The aim should be to obtain maximum accuracy and the greatest ease and efficiency of operation. Assume that standard high-performance amplifiers and multipliers, and the required power supplies are available.

The design will divide up into firstly, the layout of the operators controls, the patchboard, the number of multipliers and potentiometers, and the number and types of non-linear devices required, and, secondly, the relay circuits and switches, the cable runs and wiring, the choice of components and hardware to realize the requirements, and the layout of the parts of the computer to match the specifications decided in the first part. The internal design of the computer should aim at ease of maintenance, and simplicity and standardization of wiring. The large number of wiring joints must be made in such a manner that faults in wiring can be easily located and rectified. Such problems as lead resistance, possible earth loops and other sources of noise, temperature stability and reliability of components must be borne in mind. These and the many other problems of design give scope for a great deal of imagination and inventiveness in their solution.

REFERENCES

T. O. JEFFRIES, C. B. NEWPORT, H. A. DARKER, and R. A. FLINT, Saturn: a large analogue computing installation, *Nuclear Power*, Dec. 1961 and Jan. 1962.

J. J. GAIT, Tridac: a large analogue computer for flight simulation, *Journée int. de calcul analogique*, Sept. 1955.

chapter 5

APPLICATION OF LARGE ANALOGUE COMPUTERS

A NUCLEAR REACTOR SIMULATION—AN EXAMPLE

In order to describe the approach required when a large number of amplifiers is needed to simulate a system, an example is worked through as if it were to go on to a computer. A nuclear reactor simulation has been chosen, since, although reactor simulation studies are but a small part of the total field of application of analogue computers, the method of preparation of data and equations and the procedure on the machine is typical.

THE PHYSICAL SYSTEM

In the nuclear power stations of the Calder Hall type, heat is generated in natural uranium rods by nuclear fission. The rods are about 1 in. in diameter and 3 ft long, and are enclosed in magnesium alloy cans. The cans have various fin arrangements to improve their heat transfer properties to the coolant carbon dioxide gas. These cans of uranium fuel are stacked, one above the other, in vertical channels of about 4 in. diameter cut through graphite blocks, which are about 8 in. square and 3 ft long. The graphite blocks are built into a cylindrical stack, about 30ft high and 40 ft in diameter requiring about 40,000 blocks. The whole is enclosed in a spherical or cylindrical pressure vessel of thick steel sheet, which is connected via ducts to separate boiler vessels. The pressure vessel is filled with carbon dioxide at about 250 lb/in^2 which is driven through the graphite channels and round to the boilers by axial or centrifugal gas blowers in the boiler ducts. The gas, which is heated by the fuel elements in the channels, is used to generate steam in the boilers, which in turn is used to generate electricity in conventional turbo-generator sets.

The reactor generates heat by the process of fission. A uranium atom splits, emitting very fast moving neutrons. These fast neutrons, on average about 2·5 per fission, are very unlikely to cause another fission in natural uranium; they pass into the graphite moderator blocks and are slowed down to the thermal energy appropriate to the temperature of the graphite. These slow neutrons are very much more likely to produce fission, but only about 40 out of every 100 fast neutrons survive the slowing down process and cause fission.

When the number of fissions occurring at any instant is such that an equal number of fissions is caused as a second generation by the neutrons resulting from the first fissions, the reactor is self-sustaining, and the average rate of fission and the average neutron population remains constant. The reactor is then said to be "critical". If neutron-absorbing material is removed from the reactor core, the neutron population will increase; this increase will cause a further increase, and the rate at which fission occurs will increase more and more rapidly in an exponential manner. The reactor is then said to be "supercritical". Conversely, if absorber is inserted, the number of neutrons present and the fission rate will fall, and the reactor is subcritical. The number of neutrons available for fission from the preceding generation measures the reactivity. The natural reactivity of the reactor will be about 1·01 neutrons per neutron, and this 1 per cent supercriticality is controlled by the absorbing control rods and other effects.

The problem is to simulate the dynamic behaviour of the reactor, its control with the neutron absorbing control rods, and its cooling and safety as regards the gas flow.

APPROXIMATIONS

The complete simulation of the reactor in three dimensions is clearly impracticable, since there are about 4000 channels each containing about 7 fuel elements. The principal points of interest as regards reactor safety and output are the axial temperature and flux distributions in the channels. A steady-state solution of the equation which describes an imaginary homogeneous reactor yields a method of averaging the radial variations of flux and shows that the reactor output can be described in terms of an average channel. We therefore simplify the problem to the consideration of the simulation of one channel, with reactor flux a function only of axial position in that channel.

The fuel elements introduce axial variations into the exact solution for the flux for a homogeneous reactor, which is a cosine function. However, these variations can safely be averaged, and are small compared to some other approximations which must be made if a simulation of reasonable size is to be possible. The flux equation is thus a partial differential equation in axial distance z and in time t. Similarly, the irregularities in the temperature distribution in the radial direction in the channel, moving outwards from the centre of a fuel element, are neglected. For the simulation to be possible at all, the partial differential equations for heat transfer and for flux must be approximated by finite differences.

A further uncertainty is the behaviour of the graphite moderator. Both its thermal conductivity and its specific heat vary with time in the reactor in ways which are not fully understood, and must therefore be guessed on the basis of existing knowledge.

However, in spite of these and other approximations, which can all be justified and the errors resulting from them estimated, an extremely useful simulation can be set up.

DESCRIPTION OF THE EQUATIONS

The equation which describes the neutron behaviour, the neutron flux equation, is given as equation (1). It is a neutron balance equation saying that the rate of loss of neutrons by axial flow (in an element of length dz) plus the rate of gain of neutrons from the reactivity and the delayed neutrons gives the rate of change of the number of neutrons. The reactivity term is composed of the natural reactivity, the reactivity control, and the reactivity effects of temperature. The delayed neutrons are emitted after a certain half-life by fuel atoms which have absorbed a neutron a short time before.

The flux equation is a partial differential equation in both space and time. Since we are interested primarily in the time variations of the reactor, we split the equation into a set of finite difference equations in space. For a simulation of reasonable size, that is, below about 200 amplifiers, the flux equation is written in finite difference form as four equations giving the flux at four planes in the reactor.

At these four planes, heat-transfer equations describing radial flow of heat from the fuel and graphite to the coolant are written down. The axial flow of heat can be neglected, since the temperature gradients in the radial direction are very much greater than those in the axial direction. The flow is taken as symmetrical in the angular direction.

The heat produced by the nuclear reaction is proportional to the neutron flux. The greater part of this heat is produced in the fuel, but a small proportion appears in the graphite, can and gas. The heat-transfer equations (3) to (7) give heat balance of the form:

> Heat generated by flux, minus heat conducted to the can, plus heat conducted from the centre fuel, per second, equals rate of change of heat in the fuel surface.

The coolant equations (8), (9) give the heat balance for the channel divided into four parts centred on the four planes of the simulation. These equations are of the form:

> Heat leaving the section, minus heat entering from the preceding section in the gas, equals heat gained from the fuel can and graphite.

The Flux Equations

The basic equation can be written in the form

$$M_z^2 \frac{\partial^2 \phi}{\partial z^2} + \{k_\infty (1-\beta) - 1\} \,.\, \phi + \delta k_\infty \phi$$

$$+ k_\infty \left(\frac{\beta'}{1+D/\lambda'} + \frac{\beta''}{1+D/\lambda''} \right) . \, \phi + \frac{pS'}{\Sigma t} = l_0 D\phi, \qquad (1)$$

where ϕ = neutron flux, z measures distance along the channel,

k_∞ = reactivity constant of reactor.

δk_∞ = reactivity control, temperature and xenon effects,

$\beta = \beta' + \beta''$, M_z^2, l_0 are constants,

$$k_\infty \left(\frac{\beta'}{1+D/\lambda'} + \frac{\beta''}{1+D/\lambda''} \right) . \phi = \text{delayed neutron flux representation,}$$

$$D \equiv \frac{\mathrm{d}}{\mathrm{d}t}, \text{ and } \frac{pS'}{\Sigma t} \text{ is the neutron source term.}$$

We now write the term in $\partial^2\phi/\partial z^2$ in the form of a set of four finite difference equations. First the length of the reactor core is divided into eight equal lengths $L/2$ numbered as shown in Fig. 31 from 1 to 9. The flux at any plane i is then written as ϕ_i.

If the three-point finite difference approximation in the form $\phi_i'' = \phi_{i+1} + \phi_{i-1} - 2\phi_i$ is used, errors of up to 10 per cent can arise. Five-point Lagrange expansions about points 2, 4, 6 and 8 are therefore taken. In order to satisfy the physical boundary conditions on the value of ϕ, at planes 1 and 9, planes 0 and 10

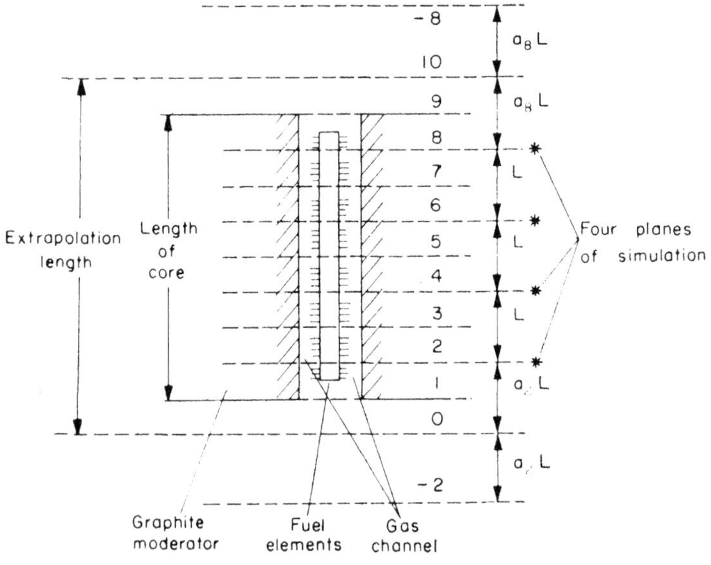

FIG. 31.

are introduced, where the flux falls to zero. These planes are a distance $a_2 L$ below plane 0 and $a_0 L$ above plane 8. Further, two planes -2 and -8, with "imaginary" fluxes of $-\phi_2$ and $-\phi_8$, are taken symmetrically below plane 0 and above plane 10. Then the flux in the neighbourhood of plane 2 is written in terms of $-\phi_2$, $\phi_0=0$, ϕ_2, ϕ_4 and ϕ_8; and near plane 4, in terms of $\phi_0=0$, ϕ_2, ϕ_4, ϕ_6 and ϕ_8; and similarly for planes 6 and 8. Using this method, it can be shown that

$$\left.\begin{aligned}
L^2\left(\frac{\partial^2\phi}{\partial z^2}\right)_2 &= -\frac{21+12a_2-4a_2^2}{2(1+3a_2+2a_2^2)}\cdot\phi_2+\frac{4a_2(3-a_2)}{1+3a_2+2a_2^2}\cdot\phi_4 \\
&\qquad -\frac{a_2(3-2a_2)}{2(2+3a_2+a_2^2)}\cdot\phi_6 \\
L^2\left(\frac{\partial^2\phi}{\partial z^2}\right)_4 &= \frac{1+3a_2}{3a_2}\cdot\phi_2-\frac{3+2a_2}{1+a_2}\cdot\phi_4+\frac{3+a_2}{2+a_2}\cdot\phi_6 \\
&\qquad -\frac{1}{3(3+a_2)}\cdot\phi_8 \\
L^2\left(\frac{\partial^2\phi}{\partial x^2}\right)_6 &= -\frac{1}{3(3+a_8)}\cdot\phi_2+\frac{3+a_8}{2+a_8}\cdot\phi_4-\frac{3+2a_8}{1+a_8}\cdot\phi_6 \\
&\qquad +\frac{1+3a_8}{3a_8}\cdot\phi_8 \\
L^2\left(\frac{\partial^2\phi}{\partial z^2}\right)_8 &= -\frac{a_8(3-2a_8)}{2(2+3a_8+a_8^2)}\cdot\phi_4+\frac{4a_8(3-a_8)}{1+3a_8+2a_8^2}\cdot\phi_6 \\
&\qquad -\frac{21+12a_8-4a_8^2}{2(1+3a_8+2a_8^2)}\,{}^{8}\phi\cdot
\end{aligned}\right\} \quad (2)$$

EXERCISE

Derive these expressions using the methods given in Chapter 1.

Heat Transfer Equations

For the fuel rod and can:

$$(v\rho c)_{f_1} DT_{f_1} = (hA)_{f_1 f_2}(T_{f_2}-T_{f_1})+S_{f_1}\phi, \quad (3)$$

$$\begin{aligned}(v\rho c)_{f_2} DT_{f_2} &= (hA)_{f_1 f_2}(T_{f_1}-T_{f_2}) \\ &\qquad -(hA)_{f,s}(T_{f_2}-T_s)+S_{f_2}\phi. \end{aligned} \quad (4)$$

For the fuel can to coolant:

$$(v\rho c)_s DT_s = (hA)_{f,s}(T_{f_2}-T_s)-(hA)_{s,c}(T_s-T_c)-1_c R. \quad (5)$$

For the graphite:

$$(v\rho c)_{m_1} DT_{m_1} = (hA)_{m_1 m_2}(T_{m_2} - T_{m_1}) - (hA)_{m_1,c}(T_{m_1} - T_c) + S_{m_1}\phi + l_c R, \quad (6)$$

$$(v\rho c)_{m_2} DT_{m_2} = (hA)_{m_1 m_2}(T_{m_1} - T_{m_2}) + S_{m_2}\phi, \quad (7)$$

where

$v\rho c$ = volume × density × specific heat (cal),
hA = heat transfer coefficient × area (cal/°C sec),
S = heat produced per unit of flux (cal/unit flux sec),
$l_c R$ = heat radiation from can to graphite (cal/sec),

and where the suffixes f_1, f_2, s, c, m_1, m_2 refer to the regions of fuel can, coolant, and moderator shown in Fig. 32, and a pair of these suffixes refers to the heat transfer between the two regions, across their common interface.

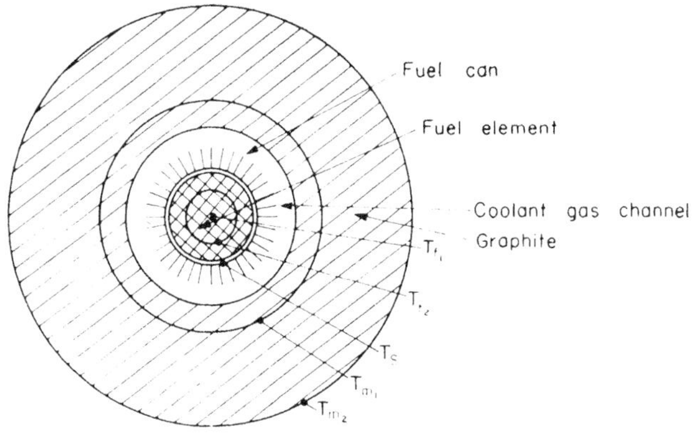

FIG. 32.

Coolant Equations

Consider the heat balance of the gas in the region of plane 2, that is, between planes 1 and 3. Using the notation of the previous section,

$$(hA)_{s,c}(T_s - T_{c_2}) + (hA)_{m,c}(T_{m_1} - T_{c_2}) = W . c_p . (T_{c_3} - T_{c_1}) + D(M . c_p . T_{c_2}), \quad (8)$$

where T_{c_1}, T_{c_2}, T_{c_3} are the coolant temperatures at planes 1, 2 and 3, W is the mass flow in g/sec, c_p the specific heat of the gas, and the mass of gas in the region is

$$M = \frac{\text{mass flow}}{\text{gas velocity}} \times \text{length of region} = Wl_c/V.$$

Assuming that the coolant behaves as a perfect gas, it is easily seen that the term $M.c_p.T_{c_2}$ is independent of temperature, since the temperature and density are inversely proportional.

We assume a linear rise of temperature from T_{c_1} to T_{c_3}; thus the equation can be written as

$$\left\{1 + \frac{(hA)_{m,c}}{2Wc_p}\right\} T_{c2} = \frac{(hA)_{s,c}}{2Wc_p}(T_s - T_{c2}) + \frac{(hA)_{m,c}}{2Wc_p} T_{m1} + T_{c1}.$$

This form is convenient, since the heat transfer depends on mass flow of gas through the channel, and can be written in the form

$$(hA)_{m,c} = (h_0 A)_{m,c}\left(\frac{W}{W_0}\right)$$

$$(hA)_{s,c} = (h_0 A)_{s,c}\, f\left(\frac{W}{W_0}\right),$$

where h_0 refers to the heat transfer coefficient at full power mass flow of W_0, and $f(W/W_0)$ is an experimental relation, approximating to a square root law.

Exercises

1. By writing the coolant equation in the form indicated, show that one function generator and a multiplier are needed in the simulation of the coolant. Draw out the simulation circuit for this equation, and for the equation in the form of (8) above, where the heat-transfer coefficients are taken to depend on mass flow as given later in the paragraph. What accuracy troubles are to be expected at low mass flow for these two methods of simulation? What cost is paid in amplifiers and equipment for a four-plane simulation to obtain the improved accuracy? (A simulation circuit is shown in Fig. 33.)

2. If T_s' and T_{m_1}' are the absolute temperatures in degrees Kelvin of the can and moderator surface, the radiation term R is

$$R = 2\pi R_{\text{fin}} \cdot \sigma(T_s'^4 - T_{m_1}'^4) \quad \text{(cal/sec unit length)},$$

where R_{fin} is about 2·5 cm, and σ is Stefan's constant $= 1{\cdot}32 \times 10^{-12}$ c.g.s. units. If T_s and T_{m_1} lie between 200° and 600°C, give circuits and scaling factors for generating $R_1 = (T_s'^4 - T_{m_1}'^4)$ using

(a) function generators,
(b) by expanding as a power series in T_s and T_{m_1},
(c) by factorizing, generating the factors and then multiplying up.

What are the faults and how accurate can each circuit be expected to be?

(*Hint*: $T_s' = T_s + t$; take $(T_s/5)$, $(T_m/5)$ and $(3R/10^{10})$ as a first try.)

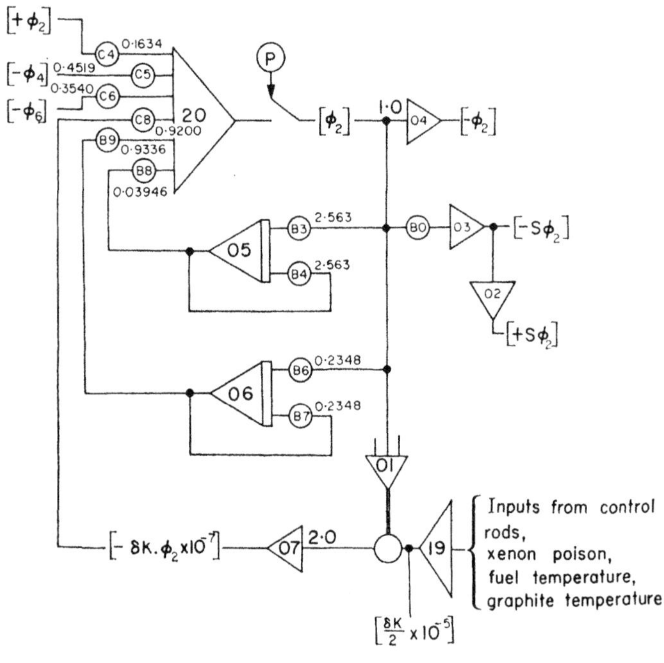

FIG. 33a.

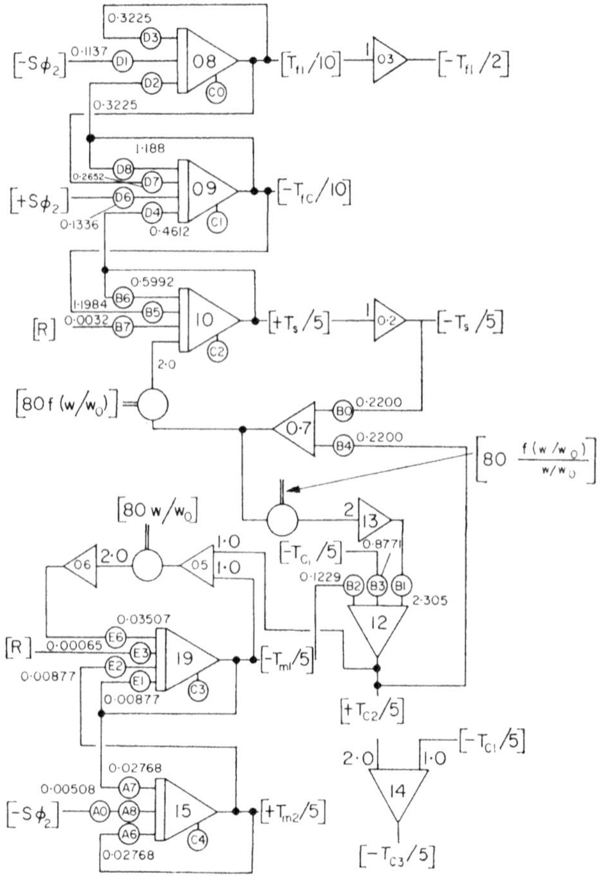
0·3225
[-Sφ2]
0·1137
08
[Tf1/10]
03
[-Tf1/2]
0·3225
1·188
0·2652
[+Sφ2]
09
[-Tfc/10]
0·1336
0·4612
0·5992
1·1984
[R]
0·0032
10
[+Ts/5]
0·2
[-Ts/5]
2·0
[80f(w/w0)]
0·7
0·2200
0·2200
[80 f(w/w0)/(w/w0)]
[80 w/w0]
2
13
06
2·0
05
1·0
1·0
[-TC1/5]
0·8771
0·1229
2·305
12
0·03507
0·00065
19
[-Tml/5]
0·00877
0·00877
[+TC2/5]
[-TC1/5]
2·0
1·0
0·02768
14
0·00508
[-Sφ2]
15
[+Tm2/5]
0·02768
[-TC3/5]

FIG. 33b.

Rack 1

Potr. No.	Gain	Input	Output
A0, A8,	0·00508	$5\phi_2$	15
A1			
A2			
A3			
A4			
A5			
A6	0·02768	15	15
A7	0·02768	19	15
A8	See A0		
A9			
B0	0·2200	02	07
B1	2·305	13	12
B2	0·1229	19	12
B3	0·8771	$\left(-\frac{(T_{c_1})}{5}\right)$	12
B4	0·2200	12	07
B5	1·1984	09	10
B6	0·5992	10	10
B7	0·0032	R	10
B8			
B9			
C0 and C9	Not used	Initial condition	
D0			
D1	0·1137	$-5\phi_2$	08
D2	0·3225	09	08
D3	0·3225	08	08
D4	0·4612	10	09
D5			
D6	0·1336	$+5\phi_2$	09
D7	0·2652	08	09
D8	1·188	09	09
D9			
E0			
E1	0·00877	19	19
E2	0·00877	15	19
E3	0·00065	R	19
E4			
E5			
E6	0·03507	06	19

Rack 2

A0–9	Not used		
B0	1·0	20	03
B1			
B2			
B3	2·563	20	05
B4	2·563	05	05
B5			
B6	0·2348	20	06
B7	0·2348	06	06
B8	0·03946	05	20
B9	0·01336	06	20

GENERAL PROCEDURE FOR PREPARING A SIMULATION

SIMULATION CIRCUITS

Once the basic forms of the equations have been fixed, and all reasonable simplifications and approximations made, sketch circuits should be drawn out. These must be simplified as much as possible, and a preliminary estimate of the amount of equipment required, together with any special circuits can be given.

The circuit for one plane of the example is given in Fig. 33.

PHYSICAL DATA

The next job in the preparation of the problem is the gathering of accurate data for all the equations. In the example, the thermal capacities of the fuel rods, moderator and can must be found. Design figures for the size of the core, the behaviour of the control rods, and the reactivity at each plane must be dug out of the engineers and physicists responsible. Decisions on which sets of data are to be used to obtain a self-consistent simulation, and how the uncertainties of the data are best handled, must be made. This is usually a lengthy and exacting job.

As soon as a rough set of values has been found, the equations can be scaled and the rough gains given. The normal procedure of guessing at maximum values and re-scaling will give finalized circuit diagrams, except for the precise values of gains.

For a large programme of work on the simulation, it is well worth while writing or having written a simple digital computer programme of the interpretive type to give the gains for the simulation from the physical data. A more complex programme will give the expected steady state and other information. As soon as accurate data are available, the machine gains and steady states can quickly be found; and when later experiments give altered thermal conductivity, heat-transfer rates and so on, the computer saves a great mass of tedious calculation. The digital programme also removes the worry of faulty arithmetic, and its printout can give the simulation gains in a neatly tabulated typescript form rather than a possibly untidy handwritten sheet of paper.

PATCHING AND SETTING UP

Once the equations have been scaled and the circuits fixed, the problem can be patched. Any special circuits should be checked on the machine or on a test computer—e.g. the simulation of the radiation from can to moderator might be checked for errors and noise.

The main layout of the problem on the patchboard should be worked out before patching starts, so that leads are kept reasonably short. As the patching proceeds, check off each amplifier, multiplier or potentiometer on the circuit diagram and equipment lists. All subsequent alterations and additions should be entered carefully on these lists, and the lists and circuits kept with the tabulated values of gains. As each section or subsection of the problem is wired on to the patchboard, check it through completely to make sure that the wires go to the right places. Cases have been known of amplifier outputs connected to earth, potentiometer inputs and sliders inverted, multiplier potentiometers connected upside-down and so on, in spite of the efforts of the patchboard designer to make things easy for the user.

PROCEDURE ON THE MACHINE

When the patchboard is fully wired up and all amplifier lists, equipment lists and circuit diagrams are ready, the problem should be set up on to the machine systematically, marking each gain as it is set. Machine faults will make themselves evident slowly—some part of the circuit will be found to integrate rapidly off to saturation due to a broken lead, or a multiplier will refuse to track round properly. It is as well to recheck the patchboard, gains and amplifier switching before calling angrily for the maintenance man. The operator who can say, "The slider of potentiometer B6 seems to have fused" and it has, is more popular with the maintenance men than the one who says, "I can't get anything from potentiometer B6—see to it" and he has forgotten to patch any volts to it. Accurate and careful work on the machine saves time for the operator, the maintenance staff, and most important saves the cost of machine time.

It should be possible to obtain some sort of steady state for new problems involving 200 amplifiers in a fortnight or so of hard work.

Once the operator has a steady state which he trusts, the programme of runs can be started. As an extension to the steady state, a standard run can be chosen and used each day to check that the machine is functioning correctly. A machine can run perfectly for weeks, and then suddenly go haywire for no apparent reason. Eventually it will be found that, for example, a capacitor unit has been changed in the course of routine maintenance, and it has been incorrectly set up, or it has an internal fault.

PROGRAMME OF WORK

The programme of runs required, sets of different gains for parameter surveys, and development work required should be prepared as the simulation is set up. Runs can then be taken from the machine at the highest rate, since everyone on the machine knows exactly what is required. Active production time, as distinct from preparation, will be divided about equally between preparing for the run, and checking up on the machine immediately prior to the run, and then taking the run. In between batches of production, gains will have to be reset and circuits changed, and a certain amount of machine idle time is almost inevitable as the runs from the machine are investigated.

SOME TYPICAL SIMULATION RESULTS

Two typical runs from a reactor simulation are shown in Figs. 34 and 35, re-plotted for clarity. If one of the four gas circulating blowers fails for some reason, the gas mass flow through the reactor will fall rapidly from 100 to about 75 per cent, whilst the reactor remains at constant power. The curves in Fig. 34 show the sector control rod movement to reduce that power, and the rise in T_{sm}, the maximum fuel can temperature. T_{sm} does not rise above about 480°C, so the sector rods move fast enough to prevent a dangerously high fuel temperature being reached. The third curve shows the variation in gas outlet temperature from the reactor.

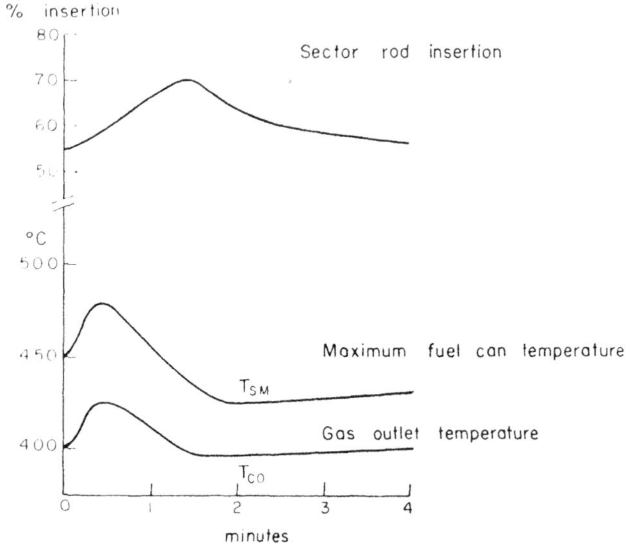

FIG. 34. Linear decrease in coolant mass flow from 100 to 70 per cent in 10 sec.

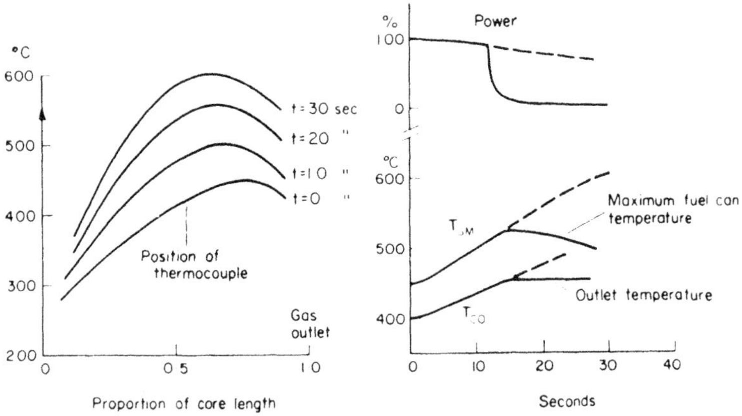

FIG. 35. Axial fuel element temperature profiles following failure of all coolant circulators.

The second set of curves, in Fig. 35, show the reactor behaviour on loss of all blower power due to voltage failure to the circulator motors. T_{sm} rises to show 515°C, and this is sufficient to trip the reactor, and causes all control rods to fall into the reactor, thus rapidly reducing power. The axial temperature profiles interpolated from the results at different times after the faults are shown in the figure.

EXAMPLES OF THE USE OF LARGE ANALOGUE COMPUTERS

THE NUCLEAR POWER INDUSTRY

An example from the nuclear industry is described in some detail earlier in this chapter. Simulations such as the one described are widely used in safety studies throughout the nuclear industry.

Other uses to which analogue techniques have been applied are studies of the build-up and decay of the fission product xenon, which has a very marked effect on the operation of the reactor as it is a strong absorber of neutrons. In a reactor operating at power, the element iodine is created as a fission product at a constant rate. The iodine then decays to xenon with a half-life of about 8 hr, and the xenon decays and is destroyed by the reactor flux to form caesium. At power, the xenon reaches a steady equilibrium level, but if the power is reduced to zero, the xenon is no longer destroyed by the flux but is still being created by decay from the iodine, and the concentration of xenon therefore increases. Xenon is a strong absorber of neutrons, and so the reactivity of the reactor is greatly reduced for a period of about 16 hr, and a return to power is often impossible. Many aspects of this "xenon poisoning" problem have been simulated, from simple simulations of the average xenon concentration following a load change to complex studies involving up to 300 amplifiers to investigate the radial stability of the flux distribution in the reactor. It is possible for a local disturbance in the flux to build up in any horizontal plane into waves of increasing amplitude, due to the time lag between the xenon and the flux. One method of solution is to write the equations in modal form, like the different modes of vibration of a stretched string. The equations to describe each mode can then be solved, and the flux obtained by summing the modes. Such studies have shown that the reactors being built in Great Britain for power stations must be controlled as a set of independent subsections, each automatically regulated by its own control rods.

A large amount of work has been done on the non-nuclear aspects of nuclear power stations, such as the heat exchangers, control rod drives and associated control loops, the electrical stability of the transformers, switching gear, motors and generators under normal and fault conditions and the simulation of a complete nuclear station in designing and developing automatic control loops for the control of the power station considered as a single unit.

In short, the nuclear industry offers a wide range of problems which are susceptible to analogue means of solution.

AIRCRAFT SIMULATION

The aircraft and missile industry makes a great deal of use of analogue computation. Apart from the extensive use of computers in the investigation of the automatic controls required to move control surfaces, vary fuel flow, steer radar and direction finding aerials, the autopilot system and many others, the actual flight dynamics are frequently simulated.

By choice of coordinate system, the equations of motion of an aircraft can be written down fairly simply from Newton's laws of motion. But the equations are seriously non-linear and complex—there are six degrees of freedom of an aircraft, three of velocity and three of rotation.

The equations are frequently linearized for small disturbances about the steady level-flight condition. The reader is probably familiar with the technique of replacing a variable x by $x_0+\delta x$, where x_0 is the steady-state value. All terms in the second or higher orders of small quantities are neglected as small, the steady-state terms cancel, and a set of equations are left which are linear, with the deviations from the steady state as variables.

A further simplification is to ignore the coupling between the rool, yaw and pitch motions about x-, y- and z-axes, taking the x-axis along the aircraft. Experiment shows that this coupling is often slight. Studies based on these simplifying assumptions are often valuable as an aid in the design of the aircraft and its associated control systems, but many refinements can be added to the basic 5 or 6 degrees of freedom simulation. The modes can be coupled and the complete large angle displacement equations, involving rotation of axes transformations in the form of direction cosine matrixes or Euler angle relationships can be simulated. The equations for the drag terms can be refined and the automatic systems and auto-pilot behaviour added, until highly complex systems result.

Stability studies and parameter surveys are two uses for such simulations. Once an aircraft is built, a simulator may be set up to match its response, and then used for training pilots, using a complete cockpit and instrumentation display system tied to the computer simulation.

MISSILE SYSTEMS

The behaviour of a guided missile system has been the subject of a great deal of analogue investigation, much of it secret. The systems for data gathering, radar searching, tracking, ground- or missile-based guidance or control, together with the airframe dynamics and the rocket motor behaviour have all been simulated, both individually and together as systems. The advantages of such simulations are typical of the advantages of the analogue approach. The behaviour of a costly, complex, ill-understood system can be evaluated prior to its construction. Optimum values of the parameters can be evaluated, and parameter surveys can be undertaken. Many of the aspects of the interrelation of the many linked servo-controls and control loops can only be evaluated on an analogue machine. Feasibility studies can be done for future systems.

A particular problem in missile systems is the response to noisy input information. The adjoint methods (Chapter 6) can be used to reduce the large number of computer runs needed to evaluate the response of a time varying system to noise. The optimum firing conditions for the minimum target "miss" distance are required under many conditions, and the adjoint method gives as output the root mean square miss distance in a single run.

A technique which has been used is to fix on a particular computer run of importance, and to devote considerable effort to obtaining a digital computer programme to give the system behaviour for that run. The comparison of the analogue simulation output and the digital answer should give added confidence in the results of the many other types of run which the simulation can produce. The established digital computer answer will serve as a check whenever any alterations or additions are made to the analogue system.

PROCESS CONTROL

The investigation of chemical plant can be very complex due to the large number of variables involved. In practice, design is very reliant on empirical methods, experience and extrapolation from previous results. Even approximate mathematical expressions for the dynamic behaviour of the plant are very hard to obtain.

Analogue computers have been used with success in many cases, and their application is increasing. If possible, the equations describing the process are written down and simulated in the ordinary way. But the amount of equipment can quickly become prohibitive. Simplifications must be made by linearizing equations, taking a coarse set of finite difference points for partial differential equations, ignoring small terms and so on.

An approach which has been used for a distillation process is to write down the large number of equations describing the evaporation and condensation on about a hundred separate plates in the distillation column and to evaluate the transfer functions relating each input to each output in turn on a digital computer. Assuming that the process is linear for a range about the design conditions, the system can be simulated by simulating these transfer functions. The resulting simulation may only involve three or four amplifiers for each transfer function, and it can then be used to study the best design of the control loops to maintain the output by varying the flows and temperatures in the system. One of the problems peculiar to chemical plant simulation is the accurate production of exponential functions which govern the reaction rates. Diode function generators can be used, but other circuits have been developed for the various special aspects of this problem. One technique is to transform all the equations into logarithmic form. This can be used to simulate systems in which a very wide range of variables must be scaled. Various automatic rescaling devices have also been tried with some success.

An important difficulty in simulating process plant is the adequate simulation of a distributed system such as a bank of heat transfer tubes in a heat exchanger. The plant parameters can be averaged, and the whole process assumed to take place at one point, or finite difference models can be set up. But an accurate representation requires a large amount of equipment, and can be difficult to manage on the computer.

Associated with such distributed systems is the difficulty of the transport delay. The length of time taken by a fluid to flow from one end of a pipe to the other cannot be simulated properly. The transfer function is of the form

$$e^{-pT} = 1 - pT + \frac{p^2T^2}{2!} - \frac{p^3T^3}{3!} + \ldots$$

The approximations are given in Fig. 24 (p. 68).

Various other methods have been tried in simulating the transport delay, using tape recorders with movable heads, scanning uniselector systems, ordinary strip chart recorders with photoelectric readers and so on. The results are seldom very good, and they are difficult to use.

In general, the advantages of the analogue computer in chemical process plant studies are those indicated previously. The standard techniques are normally sufficient for most problems, but large computers are often needed. The increasing use of analogue methods promises design and construction economies, and releases experimental plant for other work.

CONTROL ENGINEERING

Many of the techniques used in simulation on electronic analogue computers come direct from control engineering. Indeed, in some ways electronic analogue computers have been developed because of the power of the analogue in the field of control investigations. Many problems involve non-linearities and complexities far beyond the capacity of paper and pencil analysis using the most powerful tools which the studies of control theory have given the specialist. The very nature of the analogue lends itself directly to the simulation of transfer functions—the gain of an amplifier with input and feedback impedances Z_1 and Z_0 is $G(p) = -Z_0/Z_1$, and complicated impedances can be built up out of resistors and capacitors to give a wide range of transfer functions from input to output of the amplifier. (Chapter 1, Fig. 2a (p. 13), gives a short list of useful impedances.) In addition, non-linear processes can often be simulated with ease using multipliers and diode circuitry.

The advantages of speed of solution, ease of parameter alteration, ease of programming and low cost have been mentioned elsewhere in the book. Actual parts of the system involved can sometimes be incorporated for testing into an analogue investigation, e.g. a control rod and its motor can be included into a small simulation of a nuclear reactor. Simulations are often used in optimization and in feasibility studies of control systems for such reasons. The stability and accuracy of control of a system by its automatic or human operator plus automatic control can be determined in advance of its construction by the design engineer. And simulations of control systems can be used to train operating personnel. Link trainers for aircraft and the Calder Hall reactor simulator are good examples of this use.

A technique which is of wide use in control studies is that of repetitive computation. The simulation is time scaled to allow solution rates of 50 or 60c/s, and relay or other switching arranged to give a cycle of set initial conditions—compute—clamp—set initial conditions, at the 50 c/s rate. The solution is then displayed on a cathode ray tube, and, if required, photographed for record purposes. On such a simulation, the effect of altering the system parameters—loop gain, or the amount of phase advance, for example—can quickly be seen. This mode of solution is often less accurate than the conventional "real time" simulation, but has the advantage of speed and simplicity of solution.

The fields of application of analogue computers to control engineering are too great for more than a very few examples to be mentioned. Simulation has been used in steel rolling mill design, for the servos which work the rolls; in the design of feedwater controllers for boilers and reactor plant; in the control and design of surge tanks for water turbines; in the design of aircraft engine controls and of the servos to move the aircraft rudder and positional controls; and in practically every field where anything is controlled.

REFERENCES

The Calder Simulator—Calder Hall Operations School. Internal Report of the U.K.A.E.A.

R. W. HOCKNEY and T. O. JEFFRIES. The use of Analogue computers in predicting the space–time behaviour of nuclear reactors, *Proc. I.E.E.*, **109**, part A, 1961: Paper 3704.

S. M. DAVIES, Simulating Sizewell reactor transients with an Analog computer, *Nuclearonics*, May 1963.

JEROME D. KENNEDY, Dynamic business modeling, *Instruments and Control Systems*, **35**, Sept. 1962.

H. H. ROSENBROCK, The transient behaviour of distillation columns and heat exchangers—An historical and critical review, *Trans. Inst. Chem. Engrs.* **40**, 1962.

P. J. HERMANN, Simulation of steam generation in a heat exchanger, *Trans. I.R.E.*, Feb. 1962.

G. FRIEDENSOHN and D. H. SHEINGOLD, Fast time scale simulation of a reactor control system, *Trans. I.R.E.*, Aug. 1958.

GERHARD REETHOF, *Application of the Analog Computer to Predict Dynamic Performance in Typical Hydraulic Circuits*, A.S.M.E. Trans. Paper No. 57-A-97.

V. S. SOHONI and J. C. EVAN, *Electrical Analog Simulation of a Heat Exchanger Pipe Thermocouple Temperature Sensor*, A.S.M.E. Trans. Paper No. 61-WA-241, 1962.

T. I. CANNING and R. J. CHURCHILL, Some design aspects of electronic reactor simulators, *Electronic Engineering*, March 1962.

E. TORIELLI and M. CACAGNO, Dinamica dei sistemi termici a parameti distributi con particolare referimento agli scambiatori di calore della controle Nucleare di Latina, *La Termotecnica*, 26 Sept. to 1 Oct. 1961.

DONALD F. YOUNG, Electronic analogue computers for lamina flow problems, *Proc. A.S.M.E.*, Dec. 1961.

J. H. LANING and R. H. BATTIN, An application of analog computers to the statistical analysis of time-variable network, *I.R.E. Trans.*, March 1955.

J. F. MEREDITH and E. A. FREEMAN, The simulation of distributed parameter systems, with particular reference to process control problems, *Proc. I.E.E.* **105**, Part B (July 1957).

R. M. Howe and V. S. Haneman, The solution of partial differential equations by difference methods using the electronic differential analyser, *Proc. I.R.E.*, Oct. 1953.

J. M. Andrews, The dynamic storage analog computer DYSTAC, *Instruments and Control Systems*, **33**, Sept. 1960.

J. L. Dineley and E. T. Powner, Power system governor simulation, *Proc. I.E.E.*, Vol. III, Jan. 1964, No. 1.

E. L. Mitchell and G. R. Mason, In-plant application for Analog computers, *I. Soc. Am.*, No. 65–2–63.

chapter 6

GENERAL ANALOGUE DEVICES

THE DIFFERENTIAL ANALYSER

The principle of the planimeter was extended by Prof. J. Thomson, the brother of Lord Kelvin. He published a paper in the *Proceedings of the Royal Society* in 1876 "On an integrating machine having a new kinematic principle". He noted that the major fault of the planimeter was the sliding action of the wheel and, working on an idea of Prof. J. Clark Maxwell, produced the device shown diagrammatically in Fig. 36. A hardened disc can rotate about its axis, which is inclined at a slight angle to the vertical. A cylinder is mounted so that it can rotate about its axis, which is horizontal and perpendicular in direction to the axis of the disc. A ball rolls on the disc, resting against the cylinder, the whole being arranged so that the point of contact of the ball and disc lies on the horizontal line through the centre of the disc.

Suppose now that the ball, by means of a cage arrangement, is held with its centre a distance y from the centre of the disc. If the disc rotates through a small angle $\mathrm{d}x$ from its angular position x, the ball will transmit a rotation of $y\mathrm{d}x$ to the cylinder. The total rotation of the cylinder as x and y vary will then be proportional to $\int y\mathrm{d}x$.

In the same issue of the *Proceedings of the Royal Society*, Prof.

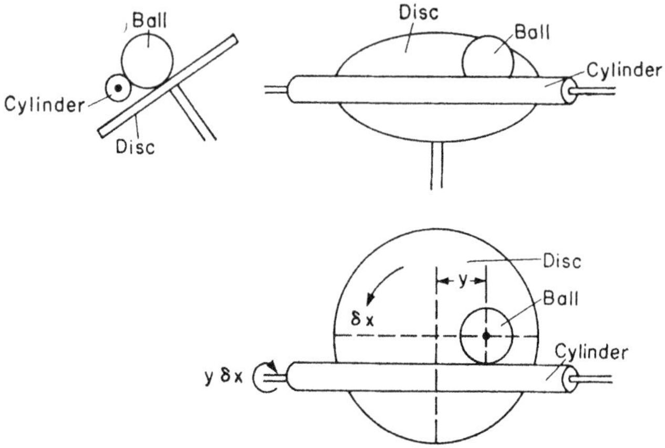

FIG. 36.

Sir W. Thomson, later Lord Kelvin, shows how to use this instrument to solve a second-order differential equation

$$\frac{\mathrm{d}}{\mathrm{d}x}\left(\frac{1}{P}\frac{\mathrm{d}u}{\mathrm{d}x}\right) = u,$$

which, quoting from a Cambridge examination question of 1874, is "easily solved" thus:

Try as a first guess the function $u = u_1$ as solution.

Form
$$u_2 = \int_0^x P\left(C - \int_0^x u_1\,\mathrm{d}x\right)\mathrm{d}x$$
$$u_3 = \int_0^x P\left(C - \int_0^x u_2\,\mathrm{d}x\right)\mathrm{d}x, \text{ etc.}$$

The series of functions u_1, u_2, $u_3, \ldots$ converge to a solution of the given differential equation.

Now Prof. Thomson formed these integrals with the aid of his brother's machine until he found a solution. "But then came a pleasant surprise. Compel agreement between the function fed into the double machine and that given out by it . . ." by establishing a connection from output to input (in modern terms a feedback connection), and "The motion of each (integrator) will thus necessarily be a solution. Thus I was led to a conclusion which was quite unexpected; and it seems to me quite remarkable that the general differential equation of the second order with variable coefficients may be rigorously, continuously, and in a single process solved by a machine."

From this pleasing conclusion Lord Kelvin proceeds in a second paper to show how to solve the general differential equation with variable coefficients using the integrating device. He envisaged its use for the solution of many problems involving vibration such as a non-uniform stretched cord, a wave in a non-uniform canal, the conduction of heat and the differential equation of the tides.

This device was not developed. However, in the 1920's V. Bush, working at M.I.T., independently developed an integrating device consisting of a wheel resting against a disc. A small hardened wheel with its axis horizontal rolls on a horizontal disc. The disc is arranged to be able to rotate about its axis and also to move on a carriage beneath the wheel in the direction of the axis of the wheel. As before, if the centre of the disc is positioned a distance y from the plane of the wheel by the carriage, and the disc rotates through an angle $\mathrm{d}x$, the wheel rotates through an angle $y\mathrm{d}x$, and so the shaft of the wheel gives an output of $\int y\mathrm{d}x$ as x and y vary.

Using this principle, a six-integrator machine was built, accurate to about 0·1 per cent, the connection between the shafts being made via gear trains.

This machine was followed during and after the war by several other machines, ranging from a two-integrator machine built out of Meccano by Prof. Hartree, and found to be accurate to better than 1·5 per cent, to a thirty-integrator machine at M.I.T., accurate to one part in 10,000 with paper tape controlled set-up and initial conditions for input, and accurate photoelectric couplings between the shafts, connected by relay circuits.

Problems such as the behaviour of emitting filaments, the control and stability of synchronous motors, and propagation along transmission lines were solved on these machines.

The torque available on the integrator output shaft is limited by the friction of the wheel against the disc; it must therefore be amplified in some way. This can be electrical, but has usually

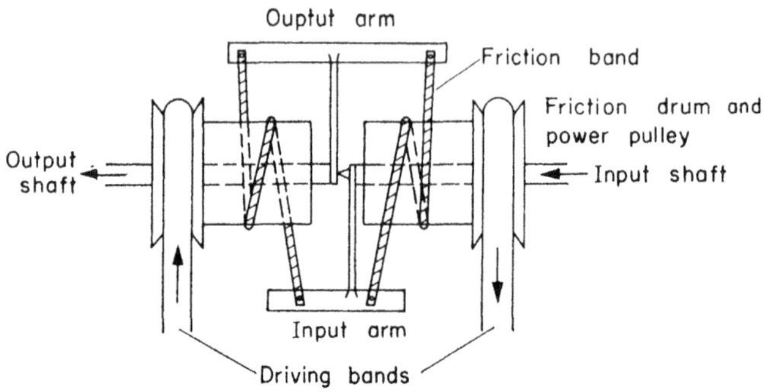

FIG. 37.

been achieved by mechanical torque amplifiers using counter-rotating drums with friction bands controlled by the input to give power to the output shaft. A study of Fig. 37 will indicate the action of this device.

SET-UP OF A DIFFERENTIAL ANALYSER

The integrator has been described; it is represented in Fig. 38a. The input to the disc position is w, and v is its rotation, giving $u = \int w \, dv$ as output. Inputs to the machine are provided from an input table by manually directing a pointer over a pre-drawn graph of the required input $q = f(p)$, a function of the variable p. Thus q might be a forcing function input to a system as a function of time, taken as p. Summation and multiplication by a constant is accomplished in gear boxes, shown in Fig. 38c. And an output table is shown in Fig. 38d; a pencil is moved across a paper according to the relationship between y and x given by the machine. A simple set-up gives the method of solution of an equation of the type

$$\frac{d^2y}{dx^2} + \mu \frac{dy}{dx} + \alpha y = F(x).$$

This is shown in Fig. 39.

Note that multiplication can be performed by two integrators together, using the relation $uv = \int u \, dv + \int v \, du$. Layouts may be derived for the generation of many functions using the property termed "regenerative connection", or, in electronic analogue terms, implicit function generation. However, the method is of much wider application, as integration can be performed with

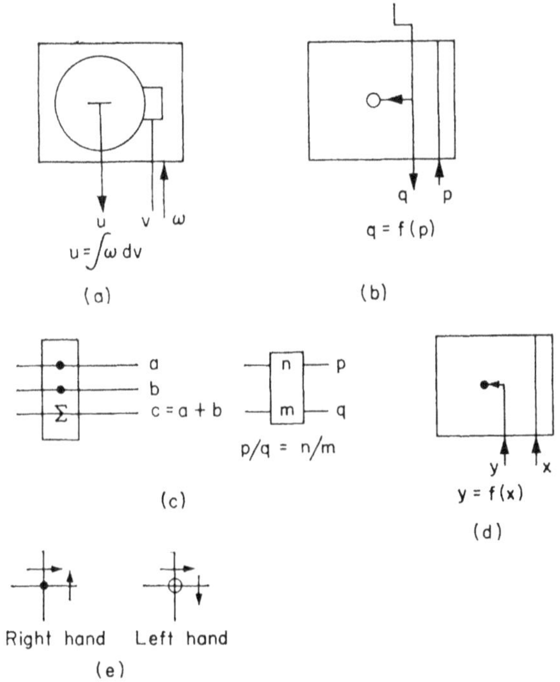

FIG. 38. (a) Integration unit. (b) Input table. (c) Summation and constant multiplication. (d) Output table. (e) Gearing between longitudinal and cross-shafts.

respect to any variable with the mechanical integrator, whereas the electronic methods of integration rely on the charging of a capacitor with time, and therefore give integration with respect to time only. This method has found an application in incremental computers, also called digital differential analysers, which are briefly described later.

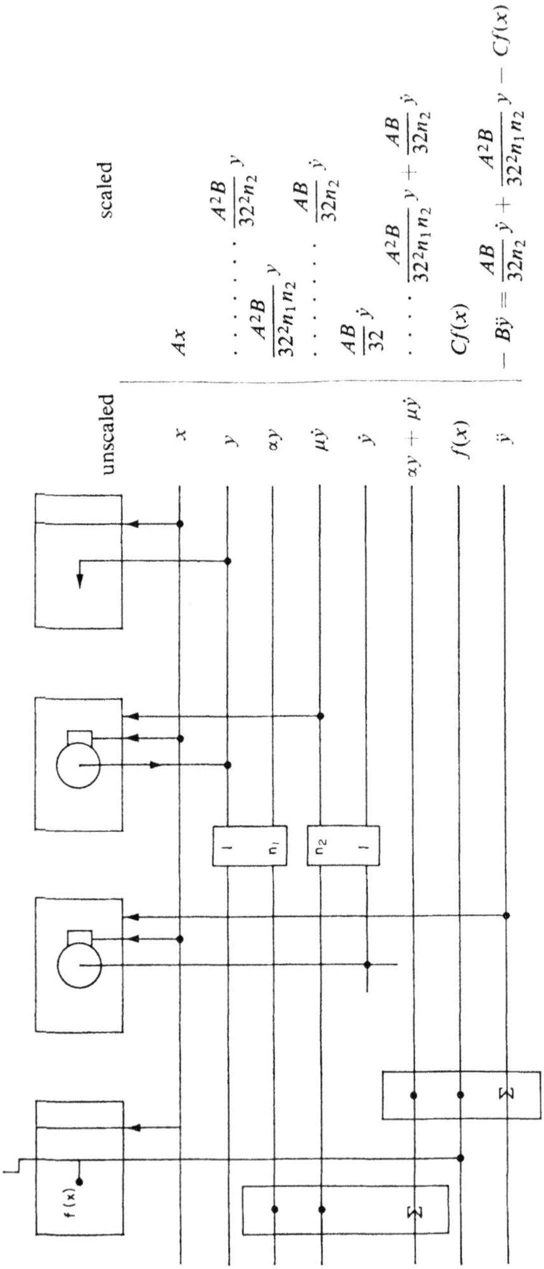

FIG. 39. Differential analyser set-up diagram for $\ddot{y} + \mu\dot{y} + \alpha y = F(x)$.

EXERCISES

1. Derive layouts to generate the functions

$$\sin\theta \text{ and } \cos\theta \text{ from } \frac{d^2y}{d\theta^2} + y = 0.$$

$$\log_e x \text{ and } \frac{1}{x} \text{ from } \int \frac{1}{x}\,dx = \log_e x,\ \frac{1}{x}\,d(\log_e x) = -\frac{1}{x}$$

$$\sqrt{x} \text{ from } \frac{dy}{dx} = \frac{1}{2x}.$$

2. Show that any gear system with an infinitely variable ratio can be used as an integrator.

THE INCREMENTAL COMPUTER

One of the advantages of the differential analyser is its ability to perform integration with respect to any variable in the computer. For some applications this facility with very great long-term stability are essential, for example in the construction of flight path and guidance-computing equipment for aeroplanes and missiles. Various aircraft firms in America and in England have built equipment for this purpose which has been referred to under the names "incremental computer" and "Digital Differential Analyser" or DDA.

Unlike a true analogue, the variables are represented by binary digits and their increments. A diagram is shown in Fig. 40a. A variable y is stored in a register Y, to which may be added its increments dy by an adding unit. Increments dx in a variable x open a gate which allows y to be added to a register R, called the remainder register. Thus y is added into R at a rate depending on the rate at which increments in x arrive, so that R accumulates a number proportional to the integral of $y.dx$. Now R will fill up progressively and will overflow. The overflow or spill digit from R can thus be used as an output $ds = k.y.dx$. Various different refinements to this basic scheme allow both signs of input to be accepted by, for example, taking an input of zero to be a subtraction and of 1 to be an addition, so that in fact a steady level in the value of a variable is represented by a train of 1's and 0's.

Improvements of technique and in the mathematical understanding of the process has enabled very high accuracy integrating units of great stability to be built. These can then be interconnected to solve differential equations, almost as if they were analogue computing units, in association with adding and other equipment. Two conventions for drawing out the integrating unit are shown in Fig.

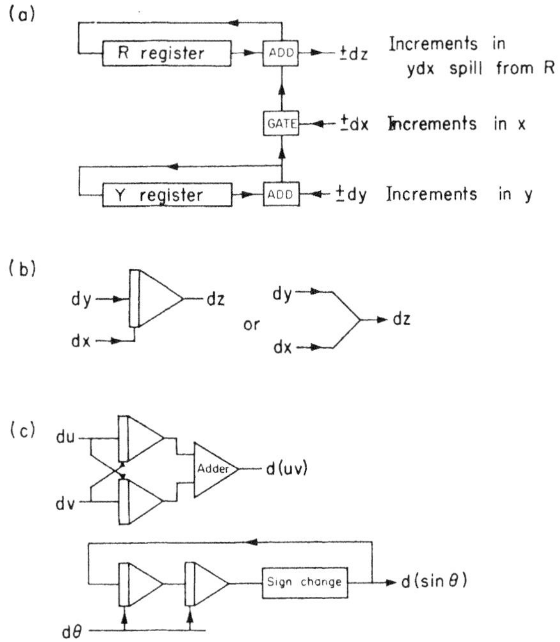

FIG. 40. (a) Block diagram of DDA element. (b) Conventional representations of DDA element. (c) Two simple DDA circuits.

40b. Since integration with respect to variables other than time is possible, the units can be interconnected in exactly similar circuits to the layouts of differential analyser equipment to give, for example, multiplication by using the fact that $d(u.v) = u.dv + v.du$, and other implicit functions as explained for the differential analyser.

A point of importance is the relation between speed of solution of a problem set-up and the accuracy of solution. For a given type of circuitry, the speed of the adding and digit transfers will be fixed. But, if the variable y is represented by a ten-binary digit register, the accuracy will be half that of an eleven-binary digit representation; however, the speed of solution (determined by the fineness or coarseness of the increments dy in the least significant digit and the fixed rate of transfer of digits) will be twice as great for the ten-digit case.

With modern techniques, incremental computing equipment can be built which give solution speeds competitive with conventional analogue equipment, with very superior accuracy, but at about twice the cost. Transistorized equipment gives great reliability and small size, and this, together with the high accuracy, have been the features particularly required in aircraft systems rather than low cost. Incremental computing equipment has no great use elsewhere in the analogue field, due to its high cost.

EXERCISE

Repeat the examples given for the differential analyser in incremental computer terms.

CONDUCTIVE ANALOGUES

The solution of Laplace's equation $\nabla^2\phi = 0$ and of Poisson's equation $\nabla^2\phi = 4\pi\rho$ in one form or another are basic problems in engineering. For complicated boundaries, the exact solution becomes impractical. One of the techniques which has been widely used in their solution is the analogue of the flow of an electric current in a conducting medium.

RESISTANCE SHEETS

A large number of engineering problems can be reduced to two-dimensional terms or to problems of cylindrical symmetry. For a two-dimensional field satisfying the Laplace equation, a uniform thin sheet of copper was used in 1845 by Kirchhoff. However, the conductivity of copper is too great for accuracy, and a convenient substance was not produced until the Western Union Telegraph Co. introduced "Teledeltos" paper in 1948. This paper is made by adding carbon black to the paper pulp during manufacture and then coating one side with a thin laquer, the other with a very thin layer of aluminium paint. The resulting paper is about 0·004 in. thick with a resistivity of the order of 20 kΩ measured from edge to edge of a square of paper of any size. Its resistivity varies by about 10 per cent in any sheet, and it is different for different directions of current flow, but it is cheap and easy to use. Boundaries may be marked by using a conducting silver paint, and it can be cut to shape with scissors and stuck to a backing plate for strength. A blunt metal sensing probe, or a ball-pointed probe can be used to measure the field pattern resulting from the boundary voltages by balancing the probe voltage in a bridge circuit. An accuracy of about 10 per cent is to be expected, which can often be within the limits of the problem data. The speed and ease of solution, even with such limited accuracy, made this a useful engineering tool.

For a solution of adequate accuracy, it is necessary that the conductive sheet is uniform in resistivity and is isotropic. Clearly local variations in resistance, systematic variations and different values in different directions will give false results, and it is these variations that are hardest to overcome satisfactorily. The electrodes used for the boundaries must have a low resistance compared to the sheet, and the input impedance of the probe must be high compared to the resistivity of the sheet or measurement inaccuracies will result. Another factor is the coefficient of resistivity with temperature, which must be low to prevent change of resistance due to the heating of the sheet from the current passing through it. The humidity of the atmosphere can also have a marked effect on the resistivity of the sheet.

Specially made conductive sheets have been used as well as the Teledeltos paper. One method is to spray a layer of conductive paint on to paper, another is to weave a cloth from alternate strands of metal and fibre. Rubber loaded with carbon can be used. One fairly accurate method is to use ordinary transparent draughtsman's paper, which appears to have no fibre structure and a very high degree of uniformity. Its conductivity appears to be due to absorbed water and salts, which form a uniform electrolyte throughout the material. The resistivity of the paper is about $3 \times 10^{10}\,\Omega$ per square, it varies widely with humidity, and it makes very carefully designed sensing equipment necessary. But conducting boundaries can be drawn on the paper with an ordinary pencil.

For the solution of Poisson's equation, the continuous source term must be approximated by discrete current sources at various points on the paper, fed with appropriate voltages from resistance chains. Another method is to use the property of superposition of the fields, and use only one source electrode, moved from point to point, and the fields resulting from each point are then added together.

The sensing probe can be very simple or highly elaborate, with lead screws to position the probe, mounted on a carriage above the paper. The probe must make a good contact with the paper without damaging it and without wearing its point flat on the paper by repeated movements across the surface of the sheet. A simple Wheatstone bridge arrangement with a meter to indicate the null can be used to plot out the equipotentials point by point. Alternatively, valve voltmeter devices or electrometer valves for high resistance sheets can be used to obtain the position of the field lines.

THE ELECTROLYTIC TANK

The electrolytic tank is a method of great power in the solution of Laplace's equation. The first such system is described by W. C. Adams in 1875, and it has been widely used since then. As in the conducting sheet analogue, a model of the system is made by enclosing in a tank of electrolyte a scale model of the boundaries of the problem. Equipotentials are made of metal, and insulating material is used for the streamlines. The model is excited by a.c. at between about 300 c/s and 10 kc/s and a probe dips into the electrolyte to measure the voltage.

It has been found that ordinary tap-water is often adequate as an electrolyte, though a low concentration of copper sulphate or of acid in distilled water has some slight advantages. As with the conductive sheet, the resistivity must be uniform, linear and of the right order of magnitude compared to the electrodes and probe. The electrolyte must be clean, it must not react chemically with the electrodes, and its surface tension should be low to minimize the effect of the meniscus at the probe tip. An a.c. excitation is chosen of a frequency which represents a compromise between the polarizing effect of d.c. at the electrodes and the increasing capacitive effects at higher frequencies.

By using a square wave excitation, both these effects can be effectively eliminated. Typically, about 20V excitation between electrodes is convenient.

For problems in two dimensions, simulating the behaviour of a medium of constant properties, a simple tank with the equipotential boundaries simulated by appropriate conductors will give the required field patterns. However, if the properties of the medium vary, for example the varying dielectric constants of the ceramic, paper and air insulation of a transformer bush, boundaries are necessary within the tank. One method is to change the depth of the tank in a step, but vertical conducting wires are needed at the step to make the current densities approaching and leaving the step uniform in the depth of the tank, whilst following the correct field lines in the horizontal plane. Another method is to represent the different media by separating different electrolytes by thin barriers with a large number of vertical wires to transmit the current from one to the other on the lines of the above. Ordinary blotting paper separators and porous sponges in the electrolyte to increase the resistivity are other methods which have been used.

For problems with cylindrical symmetry, a tilted tank can be used. A wedge-shaped quantity of electrolyte now forms the model, and the above methods of simulating changing physical properties can be used.

The simplest type of probe is a metal wire just dipped into the liquid. But this is subject to errors due to the meniscus of the liquid against the wire. Multiple arrays of probes can be used close together, and the resulting signals used to steer the probe array along the equipotentials. Various servo systems have been designed to plot the field lines automatically, but the complexity needed is not usually warranted, since the greater part of the problem time is taken in setting up and checking the problem. However, they reduce errors and tedium, and changes of the electrolyte with time have less effect on the results since the readings are taken much quicker.

With a very carefully designed system, and using great care in all measurements, accuracies of 0·1 per cent have been obtained. However, 1–2 per cent is an average figure to be expected from normal equipment.

The method has been widely used for the design of electron lens systems, for insulation problems in high voltage equipment, and in various problems of heat conduction. Some experimental work has been done to determine its possible usefulness for three-dimensional problems, but without wide application.

RESISTANCE MESHES

A method for the solution of field problems that has developed rapidly since the war is the resistance mesh. The continuous medium of the electrolytic tank or conductive sheet analogue is replaced by discrete resistances in a rectangular mesh and currents are fed in at the mesh boundary or at interior points. The voltages at the mesh nodes are read off using high impedance voltmeters, and the equipotentials can then be plotted out.

We have seen that we may approximate to the solution of the equation

$$\frac{\partial^2 f}{\partial x^2}+\frac{\partial^2 f}{\partial y^2}=0$$

by taking four equally spaced points 1, 2, 3, 4 around a fifth point number 0, and defining f_0 from f_1, f_2, f_3, and f_4 by

$$4f_0=f_1+f_2+f_3+f_4.$$

Suppose now that four points 1, 2, 3, 4, are joined to the point 0 by equal resistors R, and the voltages V_1, V_2, V_3 and V_4 are proportional to f_1, f_2, f_3 and f_4. Then we write down the current balance for the node 0 as

$$\frac{V_1-V_0}{R}+\frac{V_2-V_0}{R}+\frac{V_3-V_0}{R}+\frac{V_4-V_0}{R}=0$$

Thus

$$4V_0=V_1+V_2+V_3+V_4.$$

Thus, if the area of interest in the field problem is divided up into a rectangular mesh, and a mesh of resistors built up to link the nodes of this mesh together, when the boundary values of the problem are replaced by proportional voltages at the boundaries of the resistance mesh, the node voltages give a finite difference analogue to the problem. The procedure is as follows. First visualize a conductive sheet analogue of the original field, made of a material of resistivity ρ per square, divided up by a rectangular mesh of spacing δx and δy, shown in Fig. 41a.

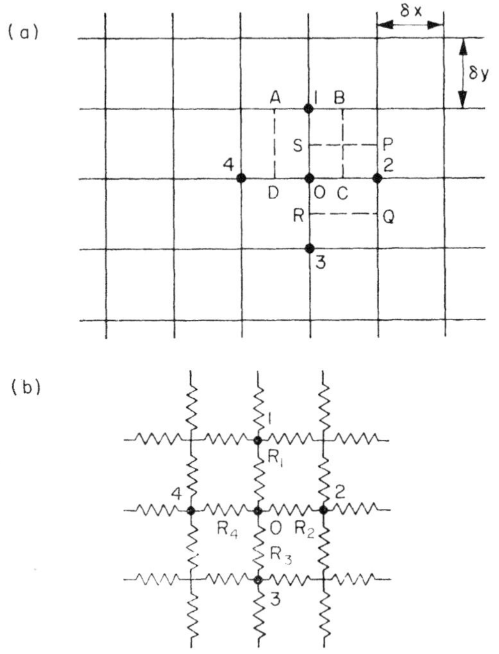

FIG. 41.

The resistance mesh analogue is shown in Fig. 41b. In this, the resistor R_1 carries current in the y-direction corresponding to the current which would flow in the sheet analogue in the y-direction between AB and CD.

So the resistance of R_1 must be proportional to the resistance between AB and CD, which is $\rho(\delta y/\delta x)$. Similarly, in the x-direction, R_2 is made proportional to the resistance between PQ and RS, that is, $\rho(\delta x/\delta y)$. Thus, we associate each resistor with a "vector area" of the field. In an exactly similar way, in three dimensions a vector volume is associated with each resistor of a three-dimensional mesh. If the mesh spacing is δx, δy, δz, this gives resistors

$$R_x = \rho \frac{\delta x}{\delta y \,.\, \delta z}, \quad R_y = \rho \frac{\delta y}{\delta z \,.\, \delta x}, \quad R_z = \rho \frac{\delta z}{\delta x \,.\, \delta y}.$$

At boundaries the resistors can be calculated in exactly the same way. For a square mesh at a boundary parallel to the mesh the vector area is half that of the other elements, so that the resistor must be twice the value of the others. Values for other boundaries are shown in Fig. 42.

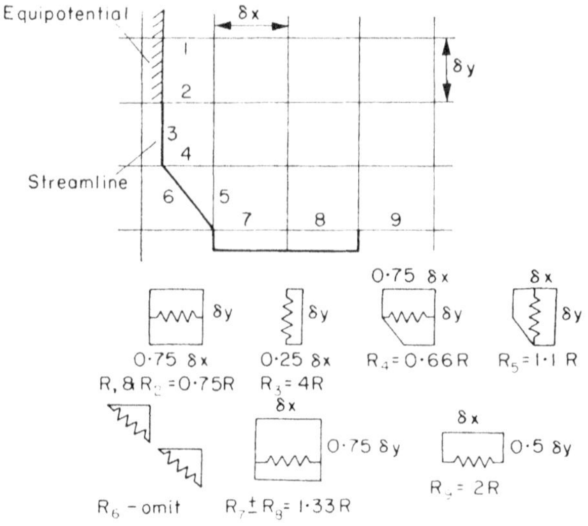

FIG. 42.

It is possible to use a finer mesh to cover areas of special interest by matching the boundaries and deriving the resistor values by the area rule as before.

EXERCISES

1. Calculate resistors for a resistance mesh analogue of the current flow across a square sheet *ABCD* from the side *AB* at voltage *V* to *CD* which is earthed, with a circular hole in the centre. If possible, set up the mesh and test it.

2. Set up a resistance mesh to study the edge effects of a parallel plate condenser.

3. Derive the resistor values for a problem in cylindrical polar co-ordinates using the vector volume method. Set up an analogue for the steady heat flow in a piston head from the flat face to a hemispherical cooling cavity, assuming no heat loss from the cylinder walls.

EXTENSIONS OF THE SIMPLE RESISTANCE MESH

A great deal of ingenuity has been used in devising methods for the solution of other types of problem to those described by the Laplace equation. For example, Poisson's equation can be simulated by feeding currents to the nodes from external sources. This can make the setting up lengthy as each time a source current is fixed, due to the resistance of the various external resistors which give the voltages for the boundary conditions, everything alters.

Dr. G. Liebmann and others have published details of various special resistance meshes for the solution of equations such as $\nabla^2\phi + k\phi = 0$ (in which the input currents to the nodes are adjusted on an iterative procedure towards the solution), for the solution of the diffusion equation $\nabla^2\phi = k\,\partial\phi/\partial t$ (by expanding as a finite difference equation in both space and time) and for the solution of the wave equation $\nabla^2\phi = k(\partial^2\phi/\partial t^2)$.

One of the most ingenious is the double mesh for the solution of problems in elasticity. The mathematical expressions defining problems in elasticity are hard to formulate and harder to solve, and the correct specification of boundary conditions can be very complex. The basic equation is the biharmonic equation

$$\nabla^4\phi = \frac{\partial^4\phi}{\partial x^4} + \frac{\partial^4\phi}{\partial x^2\,\partial y^2} + \frac{\partial^4\phi}{\partial y^2} = 0.$$

This may be solved by cascading two resistance meshes one above the other, the upper one solving for an auxiliary stress function

$$Y = \frac{\partial^2\phi}{\partial x^2} + \frac{\partial^2\phi}{\partial y^2}$$

which is specified along the boundaries of the problem and the lower solving for ϕ which can be the displacement. The nodes of the upper mesh are connected to those of the lower by resistors;

it can be shown that if these resistors are large compared to the mesh resistors, the mesh voltages form the required analogue. If resistors feed in currents to the upper mesh nodes, an analogue of $\nabla^4\phi = f$, which also arises in elasticity, is produced. Satisfactory results have been obtained from such networks using mesh resistors of about 100 Ω, interconnecting resistors of 100–200 kΩ, with the two meshes working at about 10 V and 2 V.

Dr. Redshaw and Dr. Rushton of Birmingham University have developed a negative resistance, using a simple transistor unit, to simulate various other problems in elasticity involving buckling and flexure. For the deflection ω under a tensile load P and a lateral load q, the differential equation involved is of the form

$$EI\frac{\mathrm{d}^4\omega}{\mathrm{d}x^4} - P\frac{\mathrm{d}^2\omega}{\mathrm{d}x^2} - q = 0.$$

Writing

$$M = -EI\frac{\mathrm{d}^2\omega}{\mathrm{d}x^2},$$

this becomes

$$\frac{\mathrm{d}^2M}{\mathrm{d}x^2} - \frac{P}{EI}.M + q = 0.$$

In finite difference form, this becomes

$$M_1 + M_3 - 2M_0 - \frac{Ph^2}{EI}M_0 + qh^2 = 0$$

and

$$\omega_1 + \omega_3 - 2\omega_0 + \frac{h^2}{EI}M_0 = 0.$$

These can clearly be solved by a resistance mesh for a tensile load, but a negative resistance is needed for a compression to provide for the term $(Ph^2/EI)M_0$, which becomes negative. Various problems have been solved using these negative resistance units. They are cheap, compact and easy to set up, and provide a method of overcoming the costly and cumbersome nature of inductance–capacitance networks which can also give negative impedance network elements.

Practical use of the resistance mesh has been made in many engineering problems in elasticity such as the bending of beams, the torsion of bars and hollow sections and the bending and extension of plates. Other uses are the many examples in engineering where the Laplace equation occurs, for example the flux distribution in electrical machines, electromagnetic fields in cavity resonators, forces on transformer windings and the steady state neutron flow in a nuclear reactor.

RESISTANCE–REACTANCE AND OTHER MESHES

In a precisely similar way to the set-up for resistance meshes, capacitors or inductors can be included in meshes to simulate other types of system. For example, the heat conduction equation is of the form $\nabla^2\phi = k(\partial\phi/\partial t)$. If each node of the resistance mesh simulating the space derivatives in $\nabla^2\phi$ is connected to earth by a capacitor, we simulate immediately the thermal capacity of the system and thus the whole differential equation. In one dimension this analogue is illustrated by the analogue between the equation

$$\phi_1+\phi_2-2\phi_0 = k\frac{\mathrm{d}\phi_0}{\mathrm{d}t}$$

and

$$\frac{V_1}{R}+\frac{V_2}{R}-\frac{2V_0}{R} = C\frac{\mathrm{d}V_0}{\mathrm{d}t}.$$

The reader may readily see how this may be extended to two or three dimensions, and how the rules for finding the values for resistors are extended for the capacitor values. By suitable time scaling, the simulation can be arranged for repetitive display on a cathode ray tube, or for real time computing. Values of node voltages can be recorded using high input impedance recording devices.

In a similar way, using the fact that the current in an inductor L is $(1/L)\int(V_1-V_0)\mathrm{d}t$, we may simulate the wave equation using inductance–capacitance meshes, and, by adding resistors, the modified wave equation

$$\nabla^2\phi = k_1\frac{\partial^2\phi}{\partial t^2}+k_2\frac{\partial\phi}{\partial t}+k_3\phi.$$

The understanding of the details of analogues of these two equations is left as an exercise, as it follows immediately from the above.

There are a large number of mixed inductance, resistor and capacitor mesh analogues which have been devised, but few have been used with any great success, due to the high cost, bulk, and poor accuracy of such devices in practice. In elasticity, various transformer mesh systems have been devised, but these are for the specialist. The leakage effects when using inductors, together with their internal resistance and stray capacity, limit the accuracy from practical meshes to about 5 per cent, although the mathematical theory developed by G. Kron in the course of his studies of analogues can be very powerful when a digital machine is available.

SOME SPECIAL APPLICATIONS

MONTE CARLO METHODS

The mathematical analysis of a process subject to statistical laws can be very complex in some cases. A problem of some importance in nuclear engineering is the determination of the fine structure of neutron flux in a reactor fuel element and channel. Digital methods are normally used, the machine storing the information as to the channel geometry and neutron behaviour. The machine, typically, will consider a neutron of a certain velocity, allow it to move for a random length of time, make a collision and bounce off at a random angle and repeat with another random path, until a third random number indicates that the neutron is absorbed at a collision. The numbers of paths, collisions and their locations are stored and then printed out after sufficient have been computed to provide a statistically valid result.

An analogue computer has been built to simulate this process by the English Electric Co. For a neutron of velocity V, moving in a channel with X- and Y-axes of symmetry at an angle θ to the X-axis, its position is determined by

$$x = \int_{t_0}^{t} V \cos\theta \, \mathrm{d}t + x_0,$$

$$y = \int_{t_0}^{t} V \sin\theta \, \mathrm{d}t + y_0.$$

Thus x and y can be generated by integrators from V and θ.

Since the geometry of the reactor can be built up from repeated reflections in the boundaries of the unit cell (Fig. 43), a neutron passing from one channel to the next may be simulated by reflecting it at the boundary of the cell. This may be accomplished by detecting when the X- or Y-coordinate reaches the boundary value, and then changing the sign but not the magnitude of the relevant component of velocity to reverse the direction of integration. Fast transistor switching circuits are employed for this

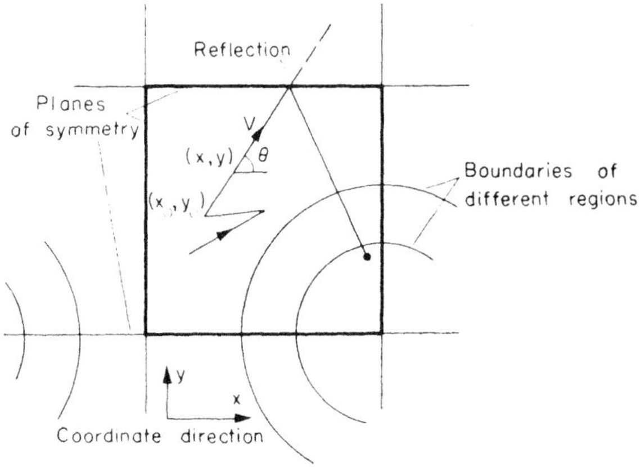

FIG. 43. Neutron path in a unit cell of a reactor lattice.

purpose, in association with standard computing amplifiers developed for the computer Saturn described on p. 80.

The random time between collisions is generated from a small radioactive source with a scintillation counter. On receipt of a pulse from the counter, the integration is stopped and an angle generator started. The angle generator produces $\sin\theta$ and $\cos\theta$, and is stopped after a time determined by a pulse from another scintillation counter, giving the direction of the reflected path. The process then repeats, until an absorption is indicated at some collision by another device. The region in which the absorption occurs is determined, and the absorption is counted automatically by decatron counters.

Many refinements can be added to simulate different properties of different regions and to detect the positions of the absorptions.

An optical technique for simulating asymmetrical configurations for shielding and other calculations has been perfected, displaying the neutron position on a cathode ray tube and detecting its position by photoelectric means. It has been found that 20–30 paths can be computed per second, which is comparable to the fastest digital computers using advanced programmes.

It is possible to use the computer to compute the complete neutron behaviour from its very fast initial velocity when born in a fission, through the complex slowing down process, to the thermal velocities at which it can cause fission. Using the computer in this way, shielding problems involving very complex geometry can be solved to a high accuracy. Due to the direct optical method of representing the geometry, problems impossible to solve on the fastest digital machines can be readily handled.

THE ADJOINT METHOD

Mathematically it may be shown that the response of a linear system to a noise input may be determined by a knowledge of the noise characteristics together with the system frequency response, or of the response of the complete system to an impulse. For a system which does not change with time, it is usually more laborious to find the frequency response experimentally than the impulse response.

However, for a system which changes with time (e.g. a rocket whose mass decreases with time), the response to a noise input depends both on the instant of observation t_2 and also on the time $t_2 - t_1$, during which the noise input has existed. Mathematically this makes both the frequency and the impulse response methods lengthy and laborious.

A technique which has been evolved to simplify the procedure makes use of a transformation of the independent variable of the differential equation which describes the system. The resulting adjoint set of equations can give a simulation which shows the system behaviour in a reversed time direction. Simulation time starts from the moment of observation and runs backwards; the simulation gives the system response at time t_2 to a noise input existing from t_1 to t_2, in terms of t_1 as independent variable, the time on the simulation.

The mathematics of this process is beyond the scope of this book, and is described in Roger and Connelly, *Analog Computation in Engineering Design*, McGraw-Hill. The following is a brief description of the method.

If the impulse response of the system is $h_{-1}(t_2, t_1)$, for observation at time t_2 following an impulse at time t_1, then $h_{-1}(t_2, t_1)$ can be non-zero only where $t_2 \geqq t_1$. That is, in the shaded region of Fig. 44a. If, then, an impulse is applied at t_1, the history of the

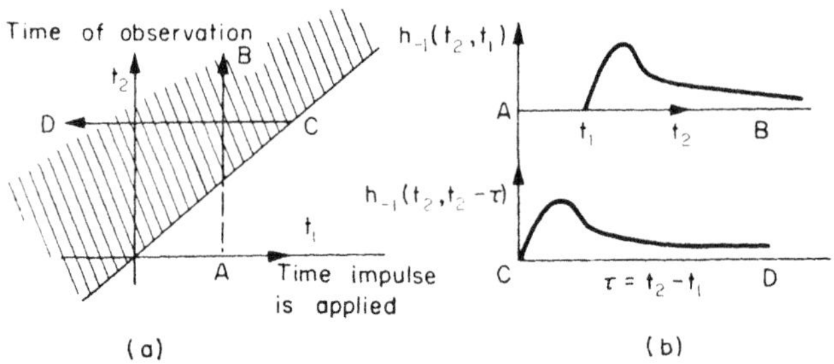

FIG. 44. Response following paths *AB* and *CD*.

system is given by following a path such as *AB*, as t_2 increases from zero (Fig. 44b). The adjoint system follows paths such as *CD*, in the direction of the arrow, where $\tau = t_2 - t_1$ is simulation time. The output which is normally required from the system is the root mean square response to noise. This is obtained by squaring and integrating the ordinate of the path *CD* with respect to τ.

Forming the Adjoint Circuit of a Simulation

Having found the adjoint simulation of a given simulation, the following rules have been shown to apply by Lanning and Battin.

1. Turn each element of the circuit round, and reverse the direction of signal flow.
2. Make all time varying parts of the system start from t_2 and run backwards in time.
3. Interchange outputs and inputs of the system. The new input will be an impulse $\delta(\tau)$ at time τ, and the output will be the impulse response $h_{-1}(t_2, t_2 - \tau)$.

4. Integrators and summers have gains of 1 only, and potentiometers have one output only. The gains may be reapportioned after the transformation.

An example given by Rogers and Connelly is:

$$\frac{1}{Kf(t)}\frac{\mathrm{d}^2\theta_0}{\mathrm{d}t^2}+\frac{1}{Kf(t)}\frac{\mathrm{d}\theta_0}{\mathrm{d}t}+\theta_0=\theta_i$$

the equation giving the output θ_0 of a time varying system to input θ_i.

The simulation circuit is shown in Fig. 45a.

Applying the above rules, the adjoint simulation is shown in Fig. 45b.

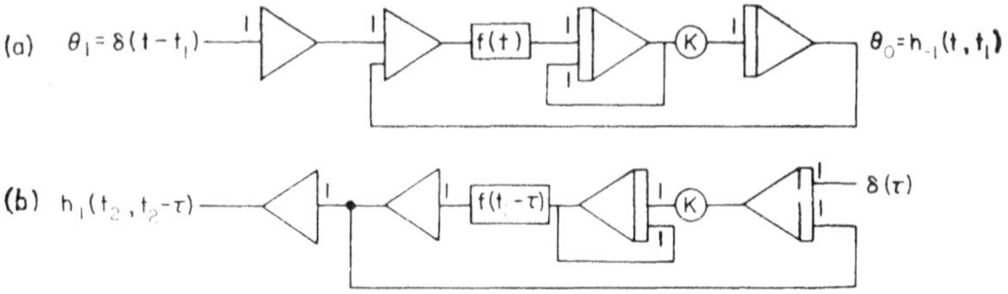

FIG. 45.

In general, any term in the original equation of the form $f_n(t)(\mathrm{d}^n/\mathrm{d}t^n)$ is replaced by a term

$$(-1)^n\frac{\mathrm{d}^n}{\mathrm{d}\tau^n}\cdot f_n(t_2-\tau),$$

and the time t of observation is replaced by t_2.

Thus the equation given above becomes

$$\frac{1}{K}\frac{\mathrm{d}^2}{\mathrm{d}\tau^2}\left\{\frac{h_{-1}(t_2,t_2-\tau)}{f(t_2-\tau)}\right\}+\frac{1}{K}\frac{\mathrm{d}}{\mathrm{d}\tau}\left\{\frac{h_{-1}(t_2,t_2-\tau)}{f(t_2-\tau)}\right\}$$
$$+h_{-1}(t_2,t_2-\tau)=\delta(\tau).$$

This is the equation solved by the adjoint circuit above.

OPERATIONAL RESEARCH

A great deal has been written about the use of analogue computers in operational research, but it seems that little practical work of importance has been done to date.

Basically, the type of problem which is solved is the determination of the optimum conditions of operation of a system, from a knowledge of its restraints expressed as inequalities. A diode simulates an inequality directly and summing amplifiers will simulate the linear relations between the variables of a linear problem.

As an example, it is desired to minimize (or maximize) a variable y (perhaps the cost of a product) where y is a linear function of variables $x_1, x_2, x_3, \ldots, x_n$ (costs of the operations required to produce the product).

Thus $$y = \alpha_1 x_1 + \alpha_2 x_2 + \ldots + \alpha_n x_n,$$

where some or all $x_i \geqq 0$ $(i = 1, 2, \ldots, n)$.

Further, there are constraints on the variables of the form

$$\begin{aligned} A_1 &= a_{11} x_1 + a_{12} x_2 + \ldots + a_{1n} x_n \leqq R_1 \\ A_2 &= a_{21} x_1 + a_{22} x_2 + \ldots + a_{2n} x_2 \leqq R_2 \\ &\cdot \quad \cdot \quad \cdot \quad \cdot \quad \cdot \quad \cdot \quad \cdot \\ A_m &= a_{m1} x_1 + a_{m1} x_2 + \ldots + a_{mn} x_n \leqq R_m \end{aligned}$$

(For example, the limits on the costs of research, development, etc.)

For constant values of α_i and a_{ij} this type of problem is called a linear programming problem.

We use a method of steepest descent—that is in this case, find the gradient of the function y as a function of the x_i and move along that direction until a constraint is reached.

Now $$\begin{aligned} y &= \frac{\partial y}{\partial x_1}\mathbf{i}_1 + \frac{\partial y}{\partial x_2}\mathbf{i}_2 + \ldots + \frac{\partial y}{\partial x_n}\mathbf{i}_n \\ &= \alpha_1 \mathbf{i}_1 + \alpha_2 \mathbf{i}_2 + \ldots + \alpha_n \mathbf{i}_n, \end{aligned}$$

where $\mathbf{i}_1, \mathbf{i}_2, \ldots, \mathbf{i}_n$ are unit vectors in the directions of the $x_1, x_2, \ldots, x_n$.

Thus, starting from any point $(x_1, x_2, \ldots, x_n)$, we must subtract (or add) increments $\alpha_1\theta$, $\alpha_2\theta$, ..., $\alpha_n\theta$ from the x-coordinates in order to optimize y. In addition, these increments must be constant with

$$A_1 \leqq R_1, A_2 \leqq R_2, \ldots, A_n \leqq R_n.$$

We therefore fix a velocity for each x_i by

$$\frac{dx_i}{dt} = \theta\alpha_i - K\sum_j a_{ji}e_j, \quad \text{summing from } j = 1 \text{ to } j = m,$$

where $e_j = A_j - R_j$ if $A_j > R_j$,

$= 0$ if $A_j \leqq R_j$.

The x_i will therefore move freely until y "runs into" a boundary defined by some R_j, then move a little beyond the boundary until

$$\alpha_i\theta = K\sum_j a_{ji}e_j,$$

and then continue along the boundary. The point therefore moves until y is optimized. In order to minimize y, θ is taken negative; to maximize, θ is taken positive. The relative values of θ and K determine the rate at which the solution is approached and the accuracy of solution. For the simulation, e_j are generated from summers, with a diode feedback to give zero output if $R_j \geqq A_j$.

The values of $Ke_1, Ke_2, \ldots, Ke_m$ and the $\alpha_i\theta$, taken as inputs to integrators, then give x_i. A diode limiter prevents the values of x_i becoming negative.

This type of technique can give solutions to problems of the nature—what is the best way of allocating labour to a job, or of setting up shops to cover a marketing area. One problem of importance is the allocation of loads to power stations on an electrical supply grid to minimize the distribution costs and to ensure that the efficient stations are used in preference to the inefficient and costly stations. It is possible to handle non-linear problems in a limited way, but the danger in all optimizing methods is that a false minimum is found before another "deeper" minimum is seen.

Related types of problem are the analysis of the rise in sales of a product following an advertising campaign, or of the personnel requirements of a firm with fluctuating sale and staff recruitment rates. It is likely that the development of these analyses will effect considerable economies in some fields.

REFERENCES

V. BUSH, The differential analyser; a new machine for solving differential equations, *J. Franklin Inst.*, 1931.

M. G. SMITH, A recent design of a mechanical analogue computer, I.Mech.E. pre-print, Jan. 1960. Symposium on Mechanical Developments in Auto-control.

J. CRANK, *The Differential Analyser*, Longmans, 1947.

J. THOMSON, On an integrating machine having a new kinematic principle, *Proc. Roy. Soc.*, 1876.

SIR W. THOMSON (LORD KELVIN), On an instrument for calculating the integral of the product of two given functions, and mechanical integration of the linear differential equation of the second order with variable coefficients, *Proc. Roy. Soc.*, 1876.

W. B. SILVER, Function generation with a DDA, *Instruments and Control Systems*, Nov. 1960.

G. C. TOOTHILL, *On the DDA*, R.A.E. Farnborough Technical Note MS-50, April 1959.

G. LIEBMANN, Resistance network analogues, *Journée int. de Calcul analogique*, Sept. 1955.

G. LIEBMANN, The solution of transient heat flow and heat transfer problems by relaxation, *Brit. J. Appl. Phys.* **6,** April 1955.

G. LIEBMANN, The solution of waveguide and cavity resonator problems with the resistance network analogue, *Proc. I.E.E.* **99,** Part IV (1952).

G. LIEBMANN, The change of airgap flux in electrical machines due to the displacement of opposed slots, *Proc. I.E.E.*, Part C, Nov. 1956, Monograph 208-M.

G. LIEBMANN, The solution of plane stress problems by an electrical analogue method, *Brit. J. Appl. Phys.* **6,** May 1955.

G. LIEBMANN, Solution of some nuclear reactor problems by the resistance network analogue method, *J. Nuclear Energy*, **2,** 213 ff. (1956).

G. LIEBMANN, Resistance network analogues with unequal meshes or subdivided meshes, *Brit. J. Appl. Phys.* **5,** Oct. 1954.

G. LIEBMANN and R. BAILEY, An improved experimental iteration method for use with resistance network analogues, *Brit. J. Appl. Phys.* **5,** Jan. 1954.

G. LIEBMANN, Solution of transient heat transfer problems by the resistance network analogue method, *Transactions A.S.M.E.*, Aug. 1956.

S. C. Redshaw, Use of an electrical analogue for the solution of a variety of torsion problems, *Brit. J. Appl. Phys.* **11,** Oct. 1960.

S. C. Redshaw and K. R. Rushton, Various electrical analogues, incorporating negative resistances, for the solution of problems in elasticity, *Brit. J. Appl. Phys.* **12,** Aug. 1961.

S. C. Redshaw and K. R. Rushton, An electrical analogue solution for the stresses near a crack or hole in a flat plate, *J. Mech. Phys. Solids*, **8,** 1960.

S. C. Redshaw and K. R. Rushton, A study of the various boundary conditions for electrical analogue solutions of the extension and flexure of flat plates, *Aeronaut. Quart.*, Aug. 1961.

H. Goldberg, The determination of electromagnetic forces in transformer windings with axial symmetry by means of a resistance network, *Annales de l'association internationale pour le calcul analogique*, No. 4, Oct. 1961.

B. E. Cunio and T. O. Jeffries, *The Determination of the Temperature Distribution in a Fuel Element by the use of a Resistance Mesh Analogue*, English Electric Internal Report W/AT 168.

W. J. Karplus, The use of electronic analogue computers with resistance network analogues, *Brit. J. Appl. Phys.* **6,** Oct. 1955.

G. Cremosnik, A. Frei, and M. J. O. Strutt, New application of impedance networks or analog computers for electronic space charge and for semiconductor diffusion problems, *Proc. I.R.E.*, May 1958.

W. J. Brignac and R. G. Swendler, Aircraft structural analysis on an analog computer, *Proc. Amer. Soc. Civil Engrs.* **86,** No. EM3 (June 1960).

Peter Liten, The string net analog of flexure of prismatic beams, *Exp. Mech.*, Nov. 1961.

K. G. Sander and J. G. Yates, The accurate mapping of electric fields in an electrolytic tank, *Proc. I.E.E.*, Part II, 1953.

K. F. Sander and J. G. Yates, A new form of electrolytic tank, *Proc. I.E.E.*, Part C, 1957.

D. McDonald, The electrolytic analogue in the design of high-voltage power transformers, *Proc. I.E.E.*, Part II, 1953.

P. E. Green, Automatic plotting of electrostatic fields, *Rev. Scientific Instruments*, **19,** 1948.

D. B. Langmuir, An automatic plotter for electron trajectories, *R.C.A. Ref.*, March 1960.

W. J. Karplus, *Analog Simulation—Solution of Field Problems*, McGraw-Hill, 1958, Chapters 5 and 6.

R. L. Dietzold, The isograph: a mechanical root-finder (untraced journal dated Dec. 1937).

L. J. Kenwood and C. H. Nicholson, *Radar Simulators*.

P. Tenger, The use of radar simulators in the Royal Navy, *Brit. I.R.E. Convention*, 29 June 1957.

R. J. Gomperts, Computing applications where analogue method appears to be superior to digital, *Brit. I.R.E. Convention*, 29 June 1957.

S. R. Sparker and J. C. Chapman, Model methods, with particular reference to three recent applications in the fields of steel, composite and concrete construction, *Structural Eng.*, March 1961.

R. E. Morris and B. Haythemthwaite, Water flow analogues for gas dynamics, *Engineering*, 19 Aug. 1960.

H. Secker and G. Gerard, Photoelastic model engineering, *Mech. Engng.*, July 1960.

R. R. M. Mallock, An electrical calculating machine, *Proc. Roy. Soc.* A, **140,** June 1933.

A. W. Hales, C.E.G.B. network analyser installations, *Electrical Times*, 3 Nov. 1960.

C. Green, H. D'Hoop, and A. Bebroux, APACHE—a breakthrough in analog computing, *I.R.E. Trans. on Electronic Computer*, Oct. 1962.

M. P. Kostenko, System development by models in U.S.S.R., *Elec. Times*, **145,** 16 Jan. 1964.

W. J. Karplus, A hybrid computer technique for treating non-linear partial differential equations, *I.E.E.E. Trans. Elec. Comp.*, EC 13–64.

I. F. Christie, The use of analogue computers for civil engineering problems, *Inst. Civil Eng. Proc.* **25,** July 1963.

INDEX

A.C. amplifier 32
Accuracy 3
See also Errors
Adjoint method 129
Aircraft industry 102
Amplifier 26ff.
chopper-stabilized 32
Analogue
computers 2, 72, 75ff.
devices 109ff.
equipment 26ff.
Applications
of analogue computers 2, 57, 61, 101
of simulators 111, 125
Approximations 88
Availability 84

Bessel functions 3, 17
Biharmonic equation 124
Boundary values 18, 90, 118ff.
Break points 44
BUSH, VANNEVAR 111

Calder Hall 86, 106
Characteristic roots 24
Chopper stabilization 32
Coefficient potentiometers 38
Coefficients, Lagrange 20
Computers
analogue 2, 4, 72, 75
digital 2, 4, 97
Conductive analogues 117ff.
Control
of amplifiers 30
of computers 73
engineering 101, 104, 105
theory 11
Co-ordinate systems
operator ∇^2 17
transformation 47, 66
Critical reaction 87
Crossed field multipliers 41
Curl operator 15

Data preparation 97
D.C. amplifiers 26
Decibel 9
Del, operator ∇ 5, 14ff.
Delay element 67, 105
Design of a computer 72
Determinants 23
Differences, finite 18

Differential analyser 3, 109, 111 ff.
Differentiation 29, 68
Digital computer 2, 4
  programmes 97
Digital Differential Analyser (D.D.A.) 113, 115
Diodes 44
Direct solution 65
Divergence operator (Div.) 15
Drift 32 ff.

Earth, virtual 26
Earthing 85
Eigenfunctions, eigenvalues 18, 24
Elasticity 124
Electrical impedance 12
Electrolytic tank 119
English Electric Co. 80, 127
Error analysis 34
Errors
  of d.c. amplifier 34
  of integrator 37
  of multipliers 40 ff.
  of summers 37, 38
Euler angles 103

Farnborough, Royal Aircraft Establishment 84
Feedback 28, 40, 110
Finite differences 18 ff, 89, 121 ff
Fission 86
Flux 15, 89, 101, 125
Fourier series 18, 24
Frequency response 10, 32, 38, 124
Function generators 44, 48

Gain 26, 29, 34, 37, 39
  setting 31, 39, 73
Generators
  function 44, 48
  noise 47

GOLDBERG, E.A. 4
Gradient operator (grad) 14

Hall effect 42
HARTREE, PROF. D. R. 4, 111
Heat conduction 18, 126
Historical introduction 3

Impedances 12
Implicit functions 65, 113
Incremental computer 113, 115
Indirect solution 65
Integration 7, 28

KELVIN, LORD KELVIN 4, 65, 109
KIRCHOFF 117
Kirchoff's laws 12, 28

Lag, simple 9
Lagrange coefficients 20
Laplace operator (*p* or *s*) 5
Laplace transforms 5, 6
  table 6
Laplace's equation 117
Limiters 44
Linear systems 8
Link trainers 106

Maintenance 74
Mathematical techniques 5
Matrices 21
Maximum values 64
Measurement 73
Missile systems 3, 103
Modes, modal solution 18, 101
Monte Carlo methods 127
Multipliers 38, 112, 116
  crossed field 41
  Hall effect 42

quarter squares 42
Reeves 43
time division 43

Negative resistance 124
Noise 38, 129
generators 47
Nuclear reactors, nuclear power 3, 86, 101

Ohm's law 12
Operational research 134
Operators
del (∇) 5, 14ff.
Laplace (*p* or *s*) 5, 6ff.
Orthogonality 24
Outage 84
Overload indication 74

Pace systems 75, 78
Parallel computers 2
Parameter surveys 99, 103
Patchboards 72, 75ff.
Patching 98
Physical data 97
Planimeter 3, 109
Poisson's equation 117, 119
Procedure
on a computer 98
for scaling 56
simulation preparation 97
Process control 104
Programme of work 99

Quarter squares multiplier 42

Reactance analogues 126
Reactors 86
Readout systems 74
Recorders 54, 74
Reeves multiplier 43
Relay controls 30, 73
Resistance
analogues 117ff.
meshes 121, 126
negative 121, 124
Resolvers 46, 66
Royal Aircraft Establishment 84

s, Laplace operator 5
Saturn computer 80
Scalars 14
Scale factors 52, 53, 56
Scaling 50ff., 56, 105
dividers 55
integrators 52
multipliers 55
summers 52
time 54
Separation of variables 17
Serial computers 2
Series, Taylor 20
Servo
multipliers 39
resolvers 46, 66
Setting up of a computer 98
Shielding calculations 129
Simulation 97, 101ff., 125
Sine–cosine potentiometers 46
Sine wave generation 69
Solartron (analogue computers) 75
Solution methods 65
Stability 11
of transformer systems 102
Steady state 97, 99
Summing
amplifier 27
junction 30

Taylor series 20
Teledeltos paper 117
THOMSON, PROF. JAMES 4, 109
Time
division multipliers 43
scaling 54

Transfer functions 7ff.
Transforms, Laplace 6
Tridac 84

Vectors 14
Virtual earth 26, 30

Wave equation 126

Xenon 101
X–Y plotters 48, 74

Zero setting 31